D1499757

Practical Design Verification

Improve design efficiency and reduce costs with this practical guide to formal and simulation-based functional verification. Giving you a theoretical and practical understanding of the key issues involved, expert authors explain both formal techniques (model checking and equivalence checking) and simulation-based techniques (coverage metrics and test generation). You get insights into practical issues including hardware verification languages (HVLs) and system-level debugging. The foundations of formal and simulation-based techniques are covered too, as are more recent research advances including transaction-level modeling and assertion-based verification, plus the theoretical underpinnings of verification, including the use of decision diagrams and Boolean satisfiability (SAT).

Dhiraj K. Pradhan is Chair of Computer Science at the University of Bristol, UK. He previously held the COE Endowed Chair Professorship in Computer Science at Texas A & M University, also serving as Founder of the Laboratory of Computer Systems there. He has also worked as a Staff Engineer at IBM, and served as the Founding CEO of Reliable Computer Technology, Inc. A Fellow of ACM, the IEEE, and the Japan Society of Promotion of Science, Professor Pradhan is the recipient of a Humboldt Prize, Germany, and has numerous major technical publications spanning more than 30 years.

Ian G. Harris is Associate Professor in the Department of Computer Science, University of California, Irvine. He is an Executive Committee Member of the IEEE Design Automation Technical Commitee (DATC) and Chair of the DATC Embedded Systems Subcommittee, as well as Chair of the IEEE Test Technology Technical Committee (TTTC) and Publicity Chair of the IEEE TTTC Tutorials and Education Group. His research interests involve the testing and validation of hardware and software systems.

Practical Design Verification

Edited by

DHIRAJ K. PRADHAN
University of Bristol, UK

IAN G. HARRIS
University of California, Irvine

CAMBRIDGE UNIVERSITY PRESS
Cambridge, New York, Melbourne, Madrid, Cape Town, Singapore, São Paulo, Delhi

Cambridge University Press
The Edinburgh Building, Cambridge CB2 8RU, UK

Published in the United States of America by Cambridge University Press, New York

www.cambridge.org
Information on this title: www.cambridge.org/9780521859721

First published 2009

Printed in the United Kingdom at the University Press, Cambridge

A catalog record for this publication is available from the British Library

Library of Congress Cataloging in Publication data
Practical design verification / Dhiraj Pradhan, Ian G. Harris.
    p.   cm.
  Includes bibliographical references and index.
  ISBN 978-0-521-85972-1 (hardback)
  1. Integrated circuits–Verification.   I. Pradhan, Diraj.   II. Harris, Ian G.   III. Title.
  TK7874.58.P73 2009
  621.3815′48–dc22
  2008050653

ISBN 978-0-521-85972-1 hardback

Contents

Contributors

Samar Abdi
Center for Embedded Computer Systems, University of California, Irvine, USA

Maciej Ciesielski
University of Massachusetts, Amherst, USA

Harry Foster
Mentor Graphics Corporation, USA

Masahiro Fujita
The University of Tokyo, Japan

Daniel Gajski
Center for Embedded Computer Systems, University of California, Irvine, USA

Ian G. Harris
University of California, Irvine, USA

Abusaleh M. Jabir
Oxford Brookes University, UK

Joao Marques-Silva
University of Southampton, UK

Dhiraj K. Pradhan
University of Bristol, UK

Matteo Sonza Reorda
Politecnico di Torino, Italy

Ernesto Sánchez
Politecnico di Torino, Italy

Giovanni Squillero
Politecnico di Torino, Italy

Shireesh Verma
Conexant Systems, Inc., USA

Wayne H. Wolf
Georgia Institute of Technology, USA

1 Model checking and equivalence checking

Masahiro Fujita

1.1 Introduction

Owing to the advances in semiconductor technology, a large and complex system that has a wide variety of functionalities has been integrated on a single chip. It is called *system-on-a-chip* (SoC) or *system LSI*, since all of the components in an electronics system are built on a single chip. Designs of SoCs are highly complicated and require many manpower-consuming processes. As a result, it has become increasingly difficult to identify all the design bugs in such a large and complex system *before* the chips are fabricated. In current designs, the verification time to check whether or not a design is correct can take 80 percent or more of the overall design time. Therefore, the development of verification techniques in each level of abstraction is indispensable.

Logic simulation is a widely used technique for the verification of a design. It simulates the output values for given input patterns. However, because the quality of simulation results deeply depends on given input patterns, there is a possibility that there exist design bugs that cannot be identified during logic simulation. Because the number of required input patterns is exponentially increased when the size of a design is increased, it is clearly impossible to verify the overall design completely by logic simulation. To solve this problem, the development of formal verification techniques is essential. In formal verification, specification and design are translated into mathematical models. Formal verification techniques verify a design by proving its correctness with mathematical reasoning, and, therefore, they can verify the overall design exhaustively. Since formal verification is a mathematical reasoning process and logic circuits compute Boolean functions, it is realized on top of basic Boolean reasoning techniques, such as binary decision diagrams (BDDs), Boolean satisfiability checking methods (so-called SAT methods), and automatic test-pattern generation techniques (ATPG) for manufacturing test fields. The performance of formal verification methods relies heavily on the performance of these techniques. Figure 1.1 shows an overview of a formal verification flow. In formal verification, both specification and design descriptions are translated into mathematical models using front-end tools. Finite state machines, temporal logic, Boolean functions, and so on, are used as mathematical models. After mathematical models are obtained, they are analyzed

Practical Design Verification, eds. Dhiraj K. Pradhan and Ian G. Harris. Published by Cambridge University Press. © Cambridge University Press 2009.

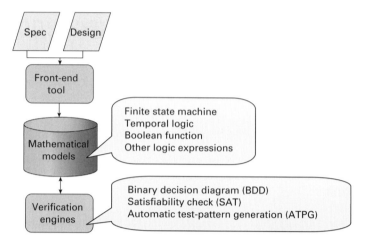

Figure 1.1   Formal verification of design descriptions

using BDD and SAT methods. Formal verification is equivalent to simulating all the cases in logic simulation. If there exists a design bug, formal verification techniques produce a counter-example to support debugging processes.

There are basically two problems in the verification of designs: model checking and equivalence checking. Model checking (or property checking) verifies whether a design satisfies the properties given as its specification. The performance of model checking has drastically improved in recent years, mainly owing to the significant progress of SAT-based efficient implementations. Equivalence checking verifies whether two given designs are equivalent or not. Equivalence checking can be applied to two designs in the same design level or in two different design levels. Depending on the types of equivalence definitions, equivalence checking can be made only on combinational parts of the circuits or on both combinational and sequential parts of the designs. In particular, the former type of equivalence checking has become very practical, and very large designs, such as those with more than 10 million gates, can be formally verified in a couple of hours.

In the actual design flow from highly abstracted design stages down to implementation levels, model checking is applied to each design level to ensure correct functionality, and equivalence checking is applied to any two different design levels so that correctness of the designs can be established. In this chapter, I first briefly review the Boolean reasoning techniques, BDD, SAT, and ATPG methods, in Section 1.2. Property checking and equivalence checking techniques are presented in Sections 1.3 and 1.4 respectively. In Section 1.5, formal verification techniques used in design levels higher than RTL are discussed.

1.2   Techniques for Boolean reasoning

In this section, I introduce three Boolean reasoning techniques, BDD, SAT, and ATPG techniques, which are the bases of formal verification methods. The performance of

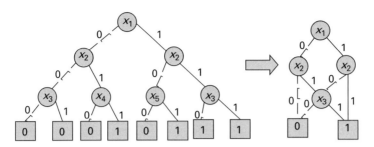

Figure 1.2   A binary decision tree representation of a Boolean function and its corresponding binary decision diagram (BDD)

formal verification methods fully relies on the performance of these techniques. In recent years SAT and ATPG methods, especially their program implementations, have been drastically improved, which make it feasible to verify real-life designs formally within reasonable time.

1.2.1   Binary decision diagrams (BDDs)

Reduced ordered binary decision diagrams (ROBDDs), simply called BDDs, are a canonical representation for Boolean functions. For many Boolean functions of practical interest in VLSI designs, BDDs provide a substantially more compact representation than other traditional alternatives, such as truth tables, sum-of-products (SOP) forms, or conjunctive normal form representations. Further, there exist efficient algorithms to manipulate BDDs. Thus, BDDs and their variants have become widely used in various areas of digital system design, including logic synthesis and formal verification of systems that can be represented in finite state machines. Binary decision diagrams represent the Boolean function as a directed acyclic graph. Let us first consider binary decision trees, an example of which appears on the left-hand side of Fig. 1.2, for the majority function, $f(x_1,x_2,x_3) = (x_1 {\wedge} x_2) {\vee} (x_2 {\wedge} x_3) {\vee} (x_1 {\wedge} x_3)$. The binary decision tree is a rooted directed tree with two kinds of node, terminal nodes and non-terminal nodes. Each non-terminal node v is labeled with a variable $var(v)$ and has two successors, $hi(v)$ and $lo(v)$, corresponding to the cases when $var(v)$ is set to 1 and 0, respectively. The edge connecting v and $hi(v)$, shown as a solid line ($lo(v)$ is shown as a dashed line), is labeled with 1 (0). Each terminal node (leaf node of the tree) is labeled by the Boolean value 0 or 1. Each truth assignment to the variables of the function has a one-to-one correspondence to a path in the tree from the root to a terminal node. This path can be traversed by starting with the root node and taking the edge corresponding to the truth value of the variable labeling the current node. The value labeling the terminal node is the value of the function under this truth assignment. This representation is, however, fairly redundant. For example, the sub-trees corresponding to the assignment ($x_1 = 0$, $x_2 = 1$) and ($x_1 = 1$, $x_2 = 0$) are isomorphic, and the vertex that corresponds to ($x_1 = 0$, $x_2 = 0$) is redundant, since both assignments to x_3 at this point have the same consequence.

A BDD could be obtained for a given Boolean function by essentially placing two restrictions on its binary decision tree representation. The first restriction imposed is a total order $<$ on the variables labeling the vertices, such that for any vertex u in the diagram, if u has a non-terminal successor v, then $var(u) < var(v)$. The second set of restrictions involves merging isomorphic sub-trees and removing redundant vertices by repeatedly applying the following three reduction rules until no further application is possible.

1. *Remove duplicate terminals* Eliminate all but one terminal vertex with a given label and redirect all arcs going to the eliminated vertices into the remaining vertex.
2. *Remove duplicate non-terminals* If two non-terminal vertices u and v have $var(u) = var(v)$, $lo(u) = lo(v)$, and $hi(u) = hi(v)$, then eliminate one of u or v and redirect all incoming arcs to the eliminated vertex to the one that remains.
3. *Remove redundant tests* If a non-terminal vertex v has $hi(v) = lo(v)$, then eliminate v and redirect all its incoming arcs to $hi(v)$.

The resulting representation is a BDD. Figure 1.2 shows an example. The graph on the right-hand side is a BDD corresponding to the binary decision tree of the majority function, shown on the left-hand side in the figure.

Binary decision diagram representations are canonical – that is, two BDDs for a given Boolean function under a given variable ordering are isomorphic. [1] Because of this the equivalence of two Boolean functions can be simply checked by a graph isomorphism check on their respective BDD representations. A function is a tautology if and only if it is isomorphic to the trivial BDD corresponding to a single terminal 1 vertex and satisfiable if and only if it is not isomorphic to the trivial 0 BDD represented by a single 0 terminal vertex. A function is independent of a variable x if and only if there is no vertex labeled with x in its BDD.

The size of a BDD representation is critically dependent on its variable order. Figure 1.3 shows two different BDD representations for the comparator function. The one on the left side uses the ordering $a_1 < a_2 < b_1 < b_2$, while the one on the right uses the order $a_1 < b_1 < a_2 < b_2$. More generally, for an n-bit comparator, the ordering $a_1 < \ldots < a_n < b_1 < \ldots < b_n$ yields a BDD with $3 \cdot 2^n - 1$ vertices, while the ordering $a_1 < b_1 < \ldots < a_n < b_n$ gives a BDD of size $3n + 2$. Thus, the size characteristics of the

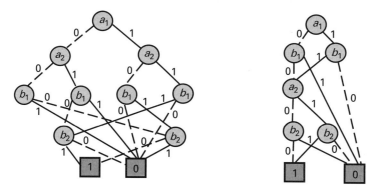

Figure 1.3   An example of how variable ordering can affect the size of an ROBDD

BDD can change from linear asymptotic growth to exponential asymptotic growth by altering the variable ordering strategy. In general, finding the optimal BDD variable order for a given function is a hard problem. Specifically, checking that a given variable order is optimal for a given function is an NP-complete problem. [2] Some classes of Boolean function are particularly difficult cases for BDDs, since any variable order results in a BDD with exponential complexity. The Boolean functions for the middle two outputs of an n-bit integer multiplier are one such example. [3]

The optimal variable order is, however, typically not necessary in order to effectively use BDDs. In practice, we need a variable order that keeps the BDD representations within reasonable limits so that suitable algorithms can manipulate them using the available computer power. In fact, many functions encountered in practical applications do have reasonably compact BDD representations. Moreover, efficient heuristics for BDD variable ordering have been developed that keep BDD sizes in check. One class of variable-ordering heuristics uses domain-specific knowledge to effect a good ordering. For example, if the Boolean function represents a logic gate network, then a depth-first traversal on the network graph can provide a good ordering. [4,5] Another technique, called *dynamic reordering* or *sifting*, [6] is an orthogonal approach, which is used when a domain-specific or constructive ordering algorithm is not available for the functions being manipulated. The technique simply performs a sequence of local reordering moves with the aim of reducing BDD size. It does this on a periodic basis to keep BDD sizes smaller and has often proved to be quite effective in practice.

One operation that is central to the construction, representation, and manipulation of BDDs is the *restriction* or *co-factoring* operation. A restriction or co-factor of f is the function that results when some variable x of f is set to a constant value k (0 or 1), denoted as $f_{x=k}$ or alternatively as f_x for $x = 1$ and $f_{\bar{x}}$ for $x = 0$. Given the two co-factors of a function, it can be expressed using the following identity, known as *Shannon's expansion*: $f = x \cdot f_x + \bar{x} \cdot f_{\bar{x}}$.

The manipulation of BDDs – that is, performing logical operations on functions represented as BDDs – is done using a single universal operation called the *ite (if-then-else)* operator (which internally makes use of the restriction operation). [7] The *ite* operator is a ternary operator, akin in functionality to a multiplexor (mux) in hardware or the *if-then-else* construct available in programming languages. It realizes the function expressed as $ite(f, g, h) = f \cdot g + \bar{f} \cdot h$, where f, g, and h are Boolean functions (possibly non-unique) represented as BDDs. In particular, *ite* can be used to implement any two-variable logic function, such as $f \oplus g = ite(f, \bar{g}, g)$ and $f \geq g = ite(f, 1, \bar{g})$.

Figure 1.4 shows the algorithm used to implement the *ite* operator for BDDs. It is a recursive algorithm where the leaves (terminal cases) of the recursion are degenerate cases of the *ite* operator for which precomputed and stored solutions are substituted, such as $ite(1,f,g) = ite(0,g,f)$ and $f\,ite(f,g,g) = g$. During the course of the algorithm, the BDD being generated may not remain fully reduced and canonical owing to the addition of new nodes, R. The *reduce*() function in the figure refers to the application of the reduction rules discussed earlier. In practical BDD packages, the need for this

```
ite(f,g,h) {
    if (terminal case) {
        return computed-result;
    } else { // general case
        let v be the top variable of (f,g,h);
        f̃← ite(f_v,g_v,h_v)
        f̃← ite(f_v,g_v,h_v)
        R = new node labeled by v
        R.hi← f̃
        R.low← g̃
        reduce(R)
        return R;
```

Figure 1.4   Algorithm to implement the *ite* operator

reduce() operation is obviated by maintaining hash tables of both unique BDD nodes and previous *ite* calls. New *ite* calls, as well as new BDD nodes (R) created through them, are looked up against these hash tables before initiating new ones, thereby dynamically maintaining and growing a reduced-ordered BDD.

1.2.2   Boolean satisfiability checker

The Boolean satisfiability (SAT) problem is a well-known constraint satisfaction problem, with many applications in the fields of VLSI computer-aided designs and artificial intelligence fields. Given a propositional formula φ, the Boolean satisfiability problem posed on φ is to determine whether there exists a variable assignment under which φ evaluates to *true*. Such an assignment, if one exists, is called a *satisfying assignment* for φ, and φ is called *satisfiable*. Otherwise, φ is said to be *unsatisfiable*. The SAT problem is known to be NP-complete. [8] However, in recent years, there have been tremendous advancements in SAT technology, making SAT solvers a viable option for solving many real-world problems.

Most SAT solvers use a *conjunctive normal form (CNF)* representation of the propositional formula. A CNF formula consists of a conjunction of clauses, each of which is a disjunction of literals, and a literal is a variable or its negation. For example $(a + b + \bar{c})(\bar{a} + c)(a + \bar{b} + c)$ is a propositional formula in CNF over the variables a, b, and c. It is composed of a conjunction of three clauses. The clause $(a + \bar{b} + c)$ is one of the clauses, a disjunction of literals a, $\bar{b}$, and c. Note that for a CNF formula to be satisfied, each of its clauses must be satisfied – that is, evaluate to *true*. There exist polynomial algorithms to transform an arbitrary propositional formula into a satisfiability equivalent CNF formula, which is satisfiable if and only if the original formula is satisfiable.

Most modern SAT solvers are based on the *Davis–Putnam–Logemann–Loveland (DPLL) procedure*. [9,10] The DPLL algorithm essentially performs a

```
sat-solve()
  if preprocess() = CONFLICT then
    return UNSAT
  while TRUE do
    if not decide-next-branch() then
      return SAT;
    while deduce() = CONFLICT do
      blevel ⇐ analyze-conflict();
      if blevel = 0 then
        return UNSAT;
      backtrack (blevel);
    done;
  done;
```

Figure 1.5   A generalized DPLL algorithm

branch-and-bound search over the space of possible Boolean assignments of the variables of the given propositional formula. It is a sound and complete algorithm – that is, it finds a satisfying assignment if and only if the given formula is satisfiable. Figure 1.5 shows the basic processing flow of the DPLL algorithm. This form provides a suitable framework for illustrating the advanced features of modern DPLL-based SAT solvers.

The first operation in the algorithm is a set of preprocessing steps (*preprocess*()) during which it may be discovered that the formula is unsatisfiable. If this is not the case, the algorithm enters the outermost loop, which consists of choosing an unassigned variable and assigning to it a value that has not been explored earlier (*decide-next-branch*()). If no such variable exists, the current partial assignment is a satisfying assignment for the formula. Otherwise, the variable assignments deducible from the current assignments are applied (*deduce*()) using a procedure known as *Boolean constraint propagation (BCP)*. This consists of an iterated application of the *unit clause rule*, which is applied on unit clauses – that is, clauses with all but one literal assigned to false and the last literal unassigned. The unit clause rule asserts the last unassigned literal of each unit clause as true, since the other assignment represents a search path that cannot lead to a satisfying assignment. A conflict occurs when a variable is asserted as true as well as false. If BCP does not lead to a conflict, the *decide-next-branch*() loop is repeated by choosing further unassigned variables and values. However, in the event of a conflict, the search backtracks (*backtrack*()) by undoing a certain number of decisions and their BCP implied assignments, based on an analysis of the conflict by *analyze-conflict*(). If all decisions need to be undone (i.e., the backtrack-level *blevel* is 0), the formula is deemed unsatisfiable, since the entire search space has been exhausted.

The original DPLL algorithm used chronological backtracking – that is, it would backtrack up to the most recent decision, for which the other value of the variable had not been tried. However, modern SAT solvers use *conflict analysis* techniques (shown as (*analyze-conflict*) in the figure) to analyze the reasons for a conflict. Conflict analysis is used to perform *conflict-driven learning* and *conflict-driven backtracking*, which were incorporated independently in the GRASP [11] and rel-sat [12] SAT solvers. Conflict-driven learning consists of adding *conflict clauses* to the formula, to avoid the same conflict in the future. Conflict-driven backtracking allows non-chronological backtracking – that is, up to the closest decision that caused the conflict. These techniques greatly improve the performance of the SAT solver on structured problems. The conflict analysis is realized using *implication graphs*, [11,13] which capture the current state of the SAT solver.

Many other advances have been made in developing the basic components that comprise the DPLL-based SAT solver: the decision engine (heuristics for choosing decision variables and values); the deduction engine (data structures and heuristics for performing BCP and detecting conflicts); and the diagnosis engine (heuristics for conflict-driven learning). [14] An interesting property of CNF representations was first exploited by Zhang in the SATO SAT solver [15] to improve the performance of BCP. It proposed the use of head and tail pointers to point to non-false literals in the list representation of a clause, and maintained the *strong invariant* that all literals before the head pointer, and all literals after the tail pointer, are false. Clearly, detection of a unit clause during BCP becomes easy – that is, when the head and tail pointers coincide on an unassigned literal. The main advantage is that the clause status is updated only when either of the head or tail literals is assigned a false value during BCP. In particular, this eliminates an update when any of the other literals in the clause is assigned a value. When the head or tail literal is assigned a false value during BCP, the associated pointer needs to be moved to another non-false literal, if it exists. This is facilitated by the strong invariant. However, during backtracking, the head or tail pointers may need to be moved back again, to maintain the strong invariant.

A different trade-off was proposed in the Chaff SAT solver. [16] Its BCP scheme, known as *two literal watching with lazy update*, is also based on tracking only two literals per clause during BCP. However, Chaff maintains a *weak invariant*, whereby the two watched literals are required to be non-false, but there is no ordering requirement with respect to other false literals. Again, detection of a unit clause during BCP is easily performed by checking whether both watched pointers coincide, and whether clause updates on assignment to other literals are eliminated.

Most of the modern-day SAT solvers incorporate the advanced techniques for conflict-based learning, branching heuristics, and efficient BCP described above as well as efficient data structures and extremely well-tuned implementations to exploit their algorithmic power fully. With these advancements, SAT solvers can now analyze formulas of up to a million variables and three to four million clauses in a few hours of runtime. Of course, these figures hold for only fairly structured SAT instances derived from certain classes of real-world problems.

1.2.3   Automatic test-pattern generation (ATPG) techniques

Automatic test-pattern generation (ATPG) is the process of generating a suite of test vectors that can be used for the purposes of testing a manufactured circuit for *manufacturing faults*. Manufacturing faults are physical defects introduced into the integrated circuit (IC), during the manufacturing process, which result in its incorrect

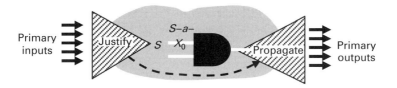

Figure 1.6   ATPG process for a single stuck-at-0 fault

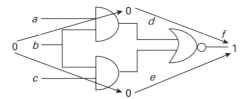

Figure 1.7   An example of implication and learning from circuits

operation. The fault we consider here is one that causes a signal to be permanently stuck at a logical value 0 or 1 (or a defect that can, for all practical purposes, be modeled as such). Such a fault is called a *stuck-at* (0 or 1) *fault*. Very efficient ATPG algorithms for stuck-at faults have been developed, which can be applied to Boolean function reasoning. Therefore, powerful formal verification techniques may be established using ATPG techniques. Thus, the purpose here is to show basic concepts and developments in ATPG so that the link of ATPG to formal verification algorithms becomes evident.

Figure 1.6 illustrates the steps involved in trying to generate a test pattern for a single stuck-at fault. In this example, the signal s is assumed to be under stuck-at-0 fault. To generate a test for s stuck-at-0, we need to find a vector of primary inputs that sets signal s to 1 (justification step) such that some primary output differs between the good circuit and the faulty circuit (propagation step).

As can be seen from the figure, the ATPG problem is basically a sort of SAT problem. We need to reason about the values of signals based on the constraints shown in the figure. Automatic test-pattern generation techniques have, however, their own historical developments rather independent from SAT method. Their algorithms and heuristics are mostly based on logic-circuit structures and properties of logic gates. This means that techniques used in ATPG methods can be used in SAT methods and vice versa.

One of the most important techniques in ATPG to speed up the test pattern generation processes is called "learning." [17,18] As seen in the previous sections, the concept of learning is also utilized in SAT methods to make them much more efficient. Similar efficiency can be achieved in ATPG processes by learning implications of values of signals from the target circuits. Figure 1.7 shows an implication example.

Suppose that input b is 0. Owing to the nature of the AND gate, d and e also become 0. This implies that f is 1. In summary, we have an implication of values that $b = 0$ implies $f = 1$. Please note that this implication process utilizes the functionality of AND and NOR gates. More learning can be made from this by using the law of contraposition, that is, we can also conclude that $f=0$ implies $b=1$. As can be seen from the figure, this learned implication is not so obvious. From $f=1$ we cannot have fixed values for d and e, since what is required from $f=1$ on d and e is that at least one of d and e must be 1, that is $d=1$ and $e=*$ (don't care) or $d=*$ and $e=1$. There are two possible values for d and e, which means that further reasoning on values of signals is not straightforward. As can be seen from the example, by using the law of contraposition, many more implications can be obtained, which will further enhance the ATPG processes.

As the discussions above on the circuits shown in Fig. 1.7, if $f=0$, there are several possible cases of values on d and e. As a result, no further simple implication of the values of signals can be made. On the other hand, in both ATPG and SAT methods, reasoning is based on case splitting and backtracking, and knowledge about necessary assignments computed from learning processes is crucial for the number of backtracks which must be performed. Backtracks occur if wrong decisions have been made, i.e., decisions considered wrong if they violate necessary assignments. Hence, it is important to realize that if *all* necessary assignments are known at every stage of the test-pattern generation process (or in general in all Boolean reasoning processes) backtracks can be avoided. Simple learning methods [17,18] cannot identify *all* necessary assignments, based, as they are, on polynomial time-complexity algorithms. The problem of identifying *all* necessary assignments is NP-complete and a method that guarantees identifying all necessary assignments must be exponential in time complexity. One such technique, which can identify all necessary assignments, is "recursive learning". [19] It involves applying learning methods in a recursive way so that even if multiple cases happen when computing implications, all such cases are exhaustively analyzed. For example, let us consider the case of $f=0$ in the circuit of Fig. 1.7. In this case, there are two cases of values for d and e, i.e., $d=1$ and $e=*$ or $d=*$ and $e=1$. Recursive learning procedures analyze one case at a time and proceed the necessary assignment analysis in a recursive way. The two cases are shown in Figs. 1.8 (a) and (b), respectively. In (a), $d=1$ implies $a=1$ and $b=1$. In (b), $e=1$ implies $b=1$ and $c=1$. The important point here is that in both cases $b=1$. That is, b is always 1. So we can conclude that $f=0$ implies $b=1$ without using the law of contraposition. In this simple example, the same implication can also be obtained by applying the law of contraposition to the implication obtained in Fig. 1.7. In general, however, much more learning can be obtained with recursive learning techniques, especially if there are more recursions. The level of recursion is defined as the number

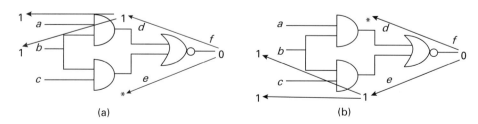

(a)   (b)

Figure 1.8   Two cases for $f=0$ in the circuit of Fig. 1.7

of nested applications of the learning techniques. In the case of Fig. 1.8, it is called recursion level 1, since only one level of case-splitting analysis is performed. For more complicated circuits, multiple levels of recursion can be applied to learn more necessary assignments. Theoretically if there are sufficiently many recursion levels, all necessary assignments can be obtained through recursive learning. On the other hand, the time for recursive learning can increase exponentially with respect to recursion levels. Therefore, there are trade-offs between the amount of learning on necessary assignments and the execution time. In practice, two levels of recursive learning give highly efficient results in terms both of test-pattern generation and its application to formal verification. As can easily be seen, similar recursion-based learning can be defined in the context of SAT methods.

As a final remark on ATPG techniques, please note that they can be applied to all problems discussed in the following sections that can be reduced to SAT problems, since ATPG and SAT methods are basically trying to solve the same problems with different viewpoints. Depending on the nature of the verification problems, sometimes ATPG-based verification is more efficient, while in other situations, SAT-based verification is better. So, in practical verification tools, both ATPG- and SAT-based methods are integrated, and they are invoked for verification with some heuristics.

1.3   Model checking techniques

Using the Boolean reasoning techniques introduced in the previous sections, I now present formal verification techniques mostly targeting hardware designs. Model checking techniques are discussed in this section and equivalence checking is discussed in the following sections.

1.3.1   Property description with temporal logic

Model checking is an automatic technique for verifying finite-state concurrent systems. The procedure involves an exhaustive search of the state space of the design to check if a given property is satisfied or not. Given sufficient computational resources, the procedure is guaranteed to terminate with a concrete yes or no answer. To apply model checking to a given system, the system needs to be expressed in a formalism amenable to model checking. Further, it is necessary to state the requirements that the system must satisfy. These requirements are typically expressed as a set of properties in a suitable logical formalism.

1.3.1.1   Kripke structures

Let AP be a set of atomic propositions. A Kripke structure over AP is a triple $M = (S, R, K)$, where

- S is a set of states,
- $R \subseteq S \times S$ is a *transition relation* that is total; that is, $(\forall s \in S)(\exists t \in S)((s,t) \in R)$,
- $K : S \to 2^{AP}$ is a *labeling function*.

A Kripke structure models the state transition graph of a Moore machine, where the outputs are functions of the current-state variables. The labeling function K associates with each state a set of atomic propositions that are true in that state. For example, in the case of a hardware system, the states S could be encoded such that there is a one-to-one mapping from S to 2^L where L is the set of latches, AP corresponds to the set of outputs of the circuits, and hence K would be a multi-output Boolean function, $K : 2^L \rightarrow 2^{AP}$ realizing the outputs.

The targets of model checking are dynamic systems, which are systems that change their states over time. *Temporal logics* are a suitable formalism for describing requirements or properties of such systems for the purpose of model checking. Temporal logics try to express system behavior over time without explicitly bringing in the notion of time. The approach used is to describe sequences of transitions between states in a dynamic system and place queries on these state sequences using a set of temporal and propositional operators allowed by the logic. Typical queries include events such as, "A particular state is *eventually* reached" or, "an erroneous scenario *never* occurs." Here we introduce one most common temporal logic, CTL, [20] which is a sub-logic of CTL*. So I first define CTL* and then show CTL.

The CTL* formulas describe properties of computation trees. Computation trees capture all possible executions of the system, starting from the initial state, and can be created by unwinding the Kripke structure into an infinite tree root at the initial state. The CTL* formulas are composed of *temporal operators* and *path quantifiers*. Path quantifiers describe the branching structure of the computation tree. There are two path quantifiers, **A** and **E**. They are applied with respect to a particular state to claim that some property is satisfied for *all computation paths* (**A**) or for *at least one computation path* (**E**) starting at the given state. Temporal operators describe the properties of a given path through the tree. There are five temporal operators in CTL*:

- **X** (*next state*) Asserts that the property is true in the next state of the path.
- **G** (*globally* or *always*) Asserts that the property is true in every state of the path.
- **F** (*eventually* or *sometime*) Requires that there exists some state on the path in which the property is true.
- **U** (*until*) This is a binary operator that holds if there exists a state on the path such that the second property holds in this state and the first property holds in each preceding state along the path.
- **R** (*release*) This is the dual of the **U** operator that asserts that the second property holds at every state along the path up to and including the first state where the first property holds. If there is no such state, then the second property should hold globally, in every state on the path.

There are two types of formula in CTL*: *state formulas* (which are true in a particular state) and *path formulas* (which are true along a specific path). If AP denotes the set of atomic propositions, the syntax of state formulas is given as follows:

- If $p \in AP$, then it is a state formula.
- If f and g are state formulas, then $\neg f$, $f \wedge g$, and $f \vee g$ are state formulas.
- If f is a path formula, then **A** f and **E** f are state formulas.

Further, path formulas are specified using the following syntax rules:

- If f is a state formula, then f is also a path formula.
- If f and g are path formulas, then $\neg f, f^{\wedge}g, f^{\vee}g, \mathbf{X}f, \mathbf{F}f, \mathbf{G}f, f\mathbf{U}g$, and $f\mathbf{R}g$ are path formulas.

We define the semantics of CTL* with respect to a Kripke structure $M=(S,R,K)$ defined earlier. An infinite sequence of states, $\psi=s_0,s_1,\ldots$, is said to be a path in M if $(\forall i. i \geq 0)((s_i,s_{i+1}) \in R)$. Let ψ^i denote the suffix of ψ starting at s_i. Let $(M,s \models f)$ denote that the state formula f is true for state s in Kripke structure M. Similarly, let $(M, \psi \models g)$ denote that the path formula g is true for path ψ in Kripke structure M. Let f_1 and f_2 be state formulas. Let g_1 and g_2 be path formulas. Then the relation $\models$ is defined inductively as follows:

- $M,s \models p$ $\quad\Leftrightarrow\quad p \in K(s)$,
- $M,s \models \neg f_1$ $\quad\Leftrightarrow\quad M,s \not\models f_1$,
- $M,s \models f_1{}^{\vee}f_2$ $\quad\Leftrightarrow\quad M,s \models f_1$ or $M,s \models f_2$,
- $M,s \models f_1{}^{\wedge}f_2$ $\quad\Leftrightarrow\quad M,s \models f_1$ and $M,s \models f_2$,
- $M,s \models \mathbf{E}g_1$ $\quad\Leftrightarrow\quad$ there exists a path ψ starting at s such that $(M, \psi \models g_1)$,
- $M,s \models \mathbf{A}g_1$ $\quad\Leftrightarrow\quad$ for every path ψ starting at s, $(M, \psi \models g_1)$,
- $M,\psi \models f_1$ $\quad\Leftrightarrow\quad s$ is the first state of ψ and $M,s \models f_1$,
- $M,\psi \models \neg g_1$ $\quad\Leftrightarrow\quad M,\psi \not\models g_1$,
- $M,\psi \models g_1{}^{\vee}g_2$ $\quad\Leftrightarrow\quad M,\psi \models g_1$ or $M,\psi \models g_2$,
- $M,\psi \models g_1{}^{\wedge}g_2$ $\quad\Leftrightarrow\quad M,\psi \models g_1$ and $M,\psi \models g_2$,
- $M,\psi \models \mathbf{X}g_1$ $\quad\Leftrightarrow\quad M,\psi^1 \models g_1$,
- $M,\psi \models \mathbf{F}g_1$ $\quad\Leftrightarrow\quad (\exists n \geq 0)(M,\psi^n \models g_1)$,
- $M,\psi \models \mathbf{G}g_1$ $\quad\Leftrightarrow\quad (\forall n \geq 0)(M,\psi^n \models g_1)$,
- $M,\psi \models g_1 \mathbf{U}g_2$ $\quad\Leftrightarrow\quad (\exists n \geq 0)((M,\psi^n \models g_2)^{\wedge}(\forall j.0 \leq j < n)(M,\psi^j \models g_2))$,
- $M,\psi \models g_1 \mathbf{R}g_2$ $\quad\Leftrightarrow\quad (\forall n \geq 0)((\forall j.0 \leq j < n)(M,\psi^j \not\models g_1) \Rightarrow (M,\psi^n \models g_2))$.

It is easily seen that the operators $^{\vee}, \neg, \mathbf{X}, \mathbf{U}$, and $\mathbf{E}$ are sufficient to express any other CTL* formula – for example, $f\mathbf{R}g \equiv \neg(\neg f\mathbf{U}\neg g), \mathbf{A}f \equiv \neg\mathbf{E}(\neg f)$, and $\mathbf{G}f \equiv \neg(\text{True }\mathbf{U}\neg f)$.

Computation tree logic (CTL) is a sub-logic of CTL*, where path formulas are restricted to be $\mathbf{X}f, \mathbf{F}f, \mathbf{G}f, f\mathbf{U}g$, and $f\mathbf{R}g$, where f and g are state formulas. There are ten basic operators in CTL – namely, $\mathbf{AX}, \mathbf{EX}, \mathbf{AG}, \mathbf{EG}, \mathbf{AF}, \mathbf{EF}, \mathbf{AR}, \mathbf{ER}, \mathbf{AU}$, and $\mathbf{EU}$. However, all ten can be expressed using the three operators $\mathbf{EX}, \mathbf{EG}$, and $\mathbf{EU}$ and using the following relationships:

- $\mathbf{AX}f \equiv \neg \mathbf{EX} \neg f$,
- $\mathbf{EF}f \equiv \mathbf{E}(\text{True } \mathbf{U}f)$,
- $\mathbf{AG}f \equiv \neg \mathbf{EF} \neg f$,
- $\mathbf{AF}f \equiv \neg \mathbf{EG} \neg f$,
- $\mathbf{A}(f\mathbf{U}g) \equiv (\neg\mathbf{E}(\neg g\mathbf{U}(\neg f^{\wedge}\neg g)))^{\wedge}(\neg\mathbf{EG}\neg g)$,
- $\mathbf{A}(f\mathbf{R}g) \equiv \neg \mathbf{E}(\neg f\mathbf{U}\neg g)$,
- $\mathbf{E}(f\mathbf{R}g) \equiv \neg \mathbf{A}(f\mathbf{U}g)$.

Properties can be broadly classified into *safety properties* and *liveness properties*. Safety properties assert that something undesirable never happens or conversely that something desirable always happens – for example, *it cannot happen that two processes are in their critical section simultaneously*, or *the message received is identical to the message sent*. On the other hand, a liveness property requires that some desirable state is repeatedly or eventually reached. Thus, liveness properties track the progress of the system and are, therefore, also referred to as *progress properties*. Examples of liveness properties are: *every bus request is eventually granted* and *a car at a traffic light is eventually allowed to pass*.

From a verification standpoint, if a system violates a safety property there will always exist a finite-length witness of that violation. Thus, safety properties can be checked on finite executions of the system. In contrast, violations of liveness properties never have finite-length witnesses. Therefore, liveness properties can only be checked on infinite-length executions of the system. In that sense, model checking of safety properties is somewhat easier than that of liveness properties.

1.3.2   Basic algorithms of CTL model checking

The model-checking problem on CTL formulas can be posed as follows:

Given a set of atomic propositions AP, a Kripke structure M = (S,R,K), a CTL formula f defined on AP, and a set of initial states I ⊆ S, does every state in I satisfy f?

The algorithm for model checking CTL formulas is an iterative procedure that computes, for each state $s \in S$, a set *label*(s) of subformulas of f that are true in s. At the start of the algorithm, that is, in the 0th iteration, each state s is labeled with the atomic propositions $K(s)$. In iteration i, subformulas of f with $i-1$ nested operators are processed, and each such subformula is added to the *label* set of the states in which it is true. Thus, upon termination, states in which f is true would have been labeled with f, and we can check if each of the initial states have been labeled with f. As discussed earlier, the CTL operators **EX**, **EU**, and **EG** and the propositional operators ¬, ∨ are sufficient to express any CTL formula. Thus, assuming that the algorithm has correctly labeled states with the subformulas f and g in iterations 0 to $i-1$, in iteration i the labeling needs to deal with the five cases, ¬f, $f \lor g$, **EX** f, **E**(f **U** g), and **EG** f. In these cases, the labeling would proceed as follows:

> **Case 1** $\varphi = \neg f$. Label *all* states, *except* those labeled with f, with the label φ.
> **Case 2** $\varphi = f \lor g$. Label all those states with label φ that have been previously labeled with either f or g.
> **Case 3** $\varphi = $ **EX** f. Label a state with φ if and only if it is a predecessor of a state labeled with f.
> **Case 4** $\varphi = $ **E**(f **U** g). Figure 1.9 shows a procedure *computeEU*(), with complexity $O(|S|+|R|)$, for handling this case. Essentially, the algorithm starts with all states labeled with g and does a backward reachability analysis from these states, using the inverse of the transition relation R, and identifying those states that have a

```
computeEU (f,g) {
  P ← {s | g ∈ label(s)}
  for all s ∈ P do
    label (s) ← label (s) ∪ {E(f U g)}
  while P ≠ ∅ do
    pick a state  s ∈ P
    P ← P − {s}
    for all  {t | R(t,s)} do
      if (E(f U g)} ∉ label (t) ∧ f ∈ label(t))
then
        label(t) ← label(t) ∪ {E(f U g)}
        P ← P ∪ {t}
    end if
```

Figure 1.9   Algorithm for labeling states of $M(S,R,K)$ that satisfy $\mathbf{E}(f \ U \ g)$

path π to the *g-labeled* states such that each state along π is labeled with f. Each of these states is then labeled with φ.

Case 5 $\varphi = \mathbf{EG}\, f$. In this case, the first step is to restrict the Kripke structure $M = (S, R, K)$ to exclude those states in which f does not hold (i.e., those not labeled by f) and restrict R and K appropriately. We construct a modified Kripke structure, $M' = (S', R', K')$ where $S' = \{s | s \in S, f \in label(s)\}$, $R' = R_{s'} \times s'$, $L' = Ls'$. With this restriction, R' may no longer be a total relation. Next, the labeling of f may be performed on M' using the following key result quoted from Clarke, Grumberg, and Peled. [20] The interested reader is referred to [20] for the proof of this result.

LEMMA 1.1 *A state s in M (S,R,K) satisfies $\varphi = EG\, f$ if and only if the following conditions hold:*

1. $s \in S'$.
2. There exists a non-trivial strongly connected component (SCC), C in the graph (S', R'), and some node $t \in C$, such that there is a path from s to t in M'.

A directed graph is called strongly connected if, for every pair of vertices u and v, there is a path from u to v and also from v to u. The *strongly connected components* (SCC) of a directed graph are its maximal strongly connected sub-graphs. These form a partition of the graph. An SCC is non-trivial if and only if it contains more than one node or it contains one only node with a self-loop. The second step in the labeling of states with $\varphi = \mathbf{EG}\, f$ is to compute the SCCs of $M' = (S', R', K')$. This can be done by Tarjan's $O(|S'|+|R'|)$ algorithm for SCC computation [21] (denoted by the function $SCC()$ in Fig. 1.10). Next, all states belonging to non-trivial SCCs are identified. This is the state set P in Figure 1.10. Finally a backward reachability search is performed from the states P, using the inverse of the transition relation R' to collect those states that have a path to some state in P such that each state along this path is labeled with f. These states are labeled with $\varphi = \mathbf{EG}\, f$. Figure 1.10 gives the

```
computeEG(f) {
  T ← {s | f∈ label(s)}
  Q ← SCC(P) // SCC computes the set of non-trivial SCCs
  of T
    P ← {s | ∃C ∈ Q, s ∈ C}
    for all s ∈ P do
      label(s) ← label(s) ∪ {EG f}
    while P ≠ ∅ do
      pick a state s ∈ P
        P ← P - {s}
      for all {t | t ∈ T ∧ R(t,s)} do
        if EG f ∉ label(t) then
          label(t) ← label(t) ∪ {EG f}
          P ← P ∪ {t}
        end if
```

Figure 1.10   Algorithm for labeling states of $M(S,R,K)$ that satisfy **EG** f

entire algorithm to perform the labeling for $\varphi = $ **EG** f. The complexity of this procedure is O($|S|+|R|$).

To summarize, the overall algorithm for model checking a CTL formula f on the Kripke structure $M = (S,R,L)$ is an iterative procedure that in each iteration picks subformulas φ of f, starting with the innermost nested subformulas and proceeding outward and labeling states that satisfy φ. Picking subformulas in this order ensures that when the algorithm processes a subformula, the labeling for all its subformulas will have been completed in earlier iterations. Thus, the labeling procedure for the current subformula amounts to solving one of the five cases discussed earlier. Each of these cases has a complexity of, at most, O($|S|+|R|$). Further, there can be, at most, $|f|$ subformulas of f and, hence, at most, as many iterations in the algorithm. This gives the overall CTL model-checking algorithm a complexity of O($|f| \cdot |S|+|R|$).

1.3.3   Symbolic model checking

Originally, model checking used an explicit representation of states. [22] A typical implementation [23] of this type of *explicit model checking* stores individual states in a large hash table, memorizing the states reached during a depth-first traversal of the state space. Since the number of states of even small systems can be very large – for example, a 128-bit shift register has 2^{128} states – this method does not scale, in particular for sequential circuits. One solution to this so-called *state explosion problem* is *symbolic model checking*, [24] which operates on sets of states instead of individual states and represents sets of states symbolically in a compact form. For the purposes of this chapter, I will limit my discussion on symbolic model checking to *simple safety properties*, also often called *invariants*, written in CTL as **AG**p. This formula specifies that, for all execution paths, globally in all states along the path, the property p holds. Alternatively, it states the property that $\neg p$, which could be some catastrophic system state, cannot be reached. Note that for finite systems, many practically relevant properties can be translated into simple safety properties. [25] Moreover, this class of property is sufficient to describe the main technologies and most common usage of symbolic model checking.

Binary decision diagrams (BDDs) and SAT methods are the two technologies primarily used to realize symbolic model-checking systems. In the following, I review symbolic model-checking techniques in the context of each of these. The field of symbolic model checking was revolutionized by the advent of binary decision diagrams. In fact, up until the relatively recent interest in SAT-based methods, symbolic model checking had been synonymous with BDD-based model checking. The paper by Bryant [26] provides a detailed discussion on representing mathematical systems such as sets and relations as Boolean functions, called *characteristic functions*, and realizing operations on these mathematical objects (sets, relations, etc.) through equivalent Boolean operations on their characteristic functions. Thus, sets and relations can be reasoned upon through BDDs by representing and manipulating their respective characteristic functions as BDDs.

The overall approach in BDD-based symbolic model checking is to represent the objects involved in model checking (essentially state sets and the transition relation of the FSM) as BDDs and realize the state traversal algorithms through suitable Boolean operations on these BDDs. The following discussion on model checking assumes a system modeled as a finite state machine (FSM). As discussed earlier, BDDs allow efficient representation of many real-life Boolean functions and efficient computation of Boolean operations on them. In particular, BDDs allow an efficient implementation of the image operation *Img*, which lies at the core of the breadth-first search in symbolic model checking. It calculates the states reachable in one step via the transition relation T from the current set of states S_C, by implicitly conjoining the BDD representing S_C with the BDD representing T and projecting the result onto the next-state variables Y (after eliminating the current-state variables X and primary input variables W).

$$Img(Y) \equiv \exists X, W \cdot S_C(X) \wedge T(X, Y, W).$$

In the context of sequential circuits, we additionally assume that the transition relation is deterministic. As shown above, however, it may depend on primary inputs, encoded by a vector W of Boolean variables, which also need to be quantified during image computation. In the terminology of program verification, *Img* calculates the strongest post condition of a given predicate. A basic algorithm for symbolic model checking simple safety properties can then be formulated as in Fig. 1.11. It represents sets of states symbolically, and searches in breadth-first order from the initial states to the bad states. Let B be the set of bad states, in which p does not hold, and I the set of initial states. This *forward model-checking* algorithm starts at the initial states and searches forward along the transition relation. In the literature, one can also find *backward model-checking* algorithms. They rely on a dual operation to the *Img* operation; *PreImg*, or equivalently the CTL operator **EX**. This calculates the set of previous states SP that may reach the given set of current states S_C in one step:

$$PreImg(X) \equiv \exists Y, W \cdot S_C(Y) \wedge T(X, Y, W).$$

```
model-check^μ_forward(I, T, B) {
  S_C ← ∅;
  S_N ← I;
  while S_C ≠ S_N do
    S_N ← S_N;
    if B ∩ S_N ≠ ∅ then
      return "found error trace to bad states";
    end if;
    S_N ← S_C ∪ Img(S_C);
  end while;
  return "no bad state reachable";
}
```

Figure 1.11   Forward least fix-point algorithm for safety properties

```
model-check^ν_backward (I, T, G) {
  S_C ← "all states";
  S_P ← G;
  while S_C ≠ S_P do
    S_C ← S_P;
    S_P ← S_C ∩ PreImg (S_C);
  end while;
  if I ⇒ S_C then
    return "only good states reachable";
  else
    return "found error trace to bad states";
  end if;
}
```

Figure 1.12   Backward greatest fix-point algorithm for safety properties

A backward model-checking algorithm can be obtained from the forward algorithm by, in essence, exchanging B with I and *Img* with *PreImg* as shown in Fig. 1.12. In practice, forward traversal is usually much faster. The reason may be that unreachable states do not have to be visited, and BDDs behave much better. However, not all temporal properties – for instance, **EX**p ^ **EX**q or **AG EX**p – can be handled with *Img* computation only. In certain cases, backward traversal is better – for instance, if the property p is an inductive invariant. In this case, the backward fix-point computation terminates after one *PreImg* computation. A general strategy is to try backward and forward traversal in parallel.

Both symbolic model-checking algorithms presented so far can be interpreted as calculating a least fix-point. [27] Significant progress has been made in both the technology and methodology of BDD-based symbolic model-checking algorithms since the first such algorithms were proposed in 1990. Current BDD-based model checkers can typically reason on systems with 200 to 400 state elements or state variables. Binary-decision-diagram-based model checking is a good match for formally verifying mission-critical properties on small to medium-sized parts or modules of a system.

Now I discuss verification methods that use SAT solvers for symbolic model checking. There are two types of method. The first set of techniques has roots in

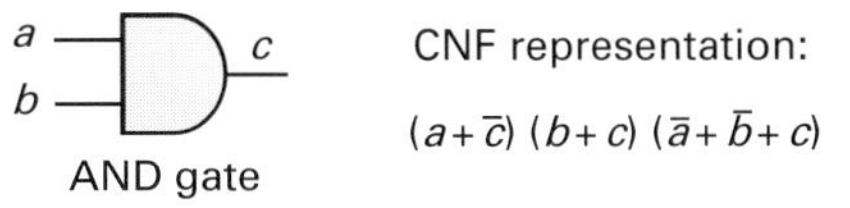

Figure 1.13 CNF representation for a logic gate

BDD-based symbolic state space search where the use of BDDs has been partially or completely replaced with SAT solvers. The second category comprises methods based on inductive reasoning. Inductive techniques are sound but usually incomplete in that they may not be able to prove every correct property. Satisfiability problems arising from Boolean circuit domains may be encoded as CNF formulas. [28] Essentially, the method involves encoding each logic gate in the circuit as a CNF formula and conjoins the CNFs generated for each gate to get the overall CNF representing the circuit. Figure 1.13 shows an example of the CNF for an AND gate. Any assertions or conditions specific to the problems can then be encoded as additional clauses and conjoined with the existing circuit CNF.

As seen above, the essential part of model checking is image operation, which forms the computational core of symbolic methods for forward model checking, as explained in the previous section.

$$S_{\mathrm{N}}(Y) = \exists X, W, Z \cdot S_{\mathrm{C}}(X) \wedge T(X, Y, W, Z).$$

In this equation, the variable sets X, Y, W, Z denote the present-state, next-state, input, and internal (needed for a CNF representation) variables, respectively; and S_{N}, S_{C}, and T denote the next states, the current states, and the transition relation, respectively. Although the designs can be encoded in CNF forms for SAT methods, the quantifiers in the image computation must be eliminated in order to apply SAT methods to the image computation. In the paper by Abdulla *et al.*, [29] the checks for property satisfaction and fix-points are formulated as SAT problems, to be solved by standard SAT solvers. The SAT problems comprise combinations of formulas S_*, representing sets of states. These are obtained by using rewriting rules to eliminate the existential quantifier in the image or pre-image operations. The most effective rule is an *inlining rule*, which substitutes an expression for a variable to be quantified; while the most expensive is rewriting the existential quantification as a disjunction, which can result in a size blow-up. A similar effort was made by Williams *et al.* [30] to use SAT solvers for CTL model checking. They use a substitution rule very effectively to eliminate the existential quantifier.

A different approach [31] can be taken, which integrates BDD-based techniques tightly into the SAT decision procedure. Here, the transition relation T is represented in CNF, and the set of reachable states S_* as BDDs. For image computation, quantifier elimination is performed by using SAT techniques to enumerate all solutions to the CNF formula, and by projecting each solution on the set of image variables (Y). The search for solutions is also constrained by the BDD for S_{P}, using a technique called *BDD bounding*, whereby any partial solution in SAT that is inconsistent with the BDD is regarded as a conflict. This technique is also used effectively to avoid repeating image-set solutions by bounding against the current S_N. In this approach, BDD-based sub-problems are also generated on the fly, under a partially explored path in SAT. Though this procedure can be used to perform cube enumeration in SAT alone, the use of BDD sub-problems is highly beneficial in handling large designs.

Inductive reasoning is another way to implement model checking with SAT methods. The inductive proof for verifying a property $P = \mathbf{AG}p$ can be derived using a SAT solver by checking the formulas φ_{base} (the base case) and φ_{induc} (the induction step) for unsatisfiability.

$$\varphi_{\text{base}} = I \wedge \neg P_0,$$
$$\varphi_{\text{induc}} = P_k \wedge T(k, k+1) \wedge (\neg P_{k+1}).$$

If φ_{induc} is unsatisfiable, the property P is called an *inductive invariant*. Both formulas, if unsatisfiable, provide a sufficient (but not necessary) condition for verifying P. However, the above form of induction, known as *simple induction*, is not powerful enough to verify many properties. More powerful forms of induction, known as *induction with depth* and *unique states induction*, are proposed to verify safety properties. For induction with depth n, the above formulas become:

$$\varphi_{\text{base}}^n = I \wedge \left[\bigwedge_{i=0}^{n-1} T(i, i+1) \right] \wedge \bigvee_{i=0}^{n} \neg P_i,$$

$$\varphi_{\text{induc}}^n = \left[\bigwedge_{j=k}^{k+n} P_j \right] \wedge \left[\bigwedge_{i=k}^{k+n} T(i, i+1) \right] \wedge \neg P_{k+n+1}.$$

Essentially, induction with depth corresponds to strengthening the induction hypothesis, by imposing the original induction hypothesis on n consecutive time-frames. This can be further strengthened by requiring that the states appearing on each time-frame be unique (*unique states induction*). This restriction results in a complete method for simple safety properties.

1.3.4   Practical model checking

It is a well-recognized fact that traditional simulation methods, while quite efficient and scalable, are unable to provide the validation coverage needed to uncover difficult corner-case bugs. Formal verification techniques can potentially provide complete coverage. However, the current state of the art in formal methods cannot handle the complexity and size of modern-day VLSI designs. Thus, there have been significant efforts in the development of semi-formal validation technologies that attempt to provide the scalability of simulation techniques and the coverage of formal verification. One such attempt, bounded model checking based on SAT solvers, has made dramatic progress. Bounded model checking based on SAT methods was introduced in [32] and is rapidly gaining popularity as a complementary technique to BDD-based symbolic model checking. Given a temporal logic property P to be verified on a finite transition system M, the essential idea is to search for counter-examples in the space of all executions of M whose length is bounded by some integer k.

The problem is formulated by constructing the following propositional formula:

$$\varphi^k = I \wedge \bigcap_{i=0}^{k-1} T_i \wedge (\neg P^k), \tag{1.1}$$

where I is the characteristic function for the set of initial states of M, T_i is the characteristic function of the transition relation of M for time step i. Thus, the formula $I \wedge \bigcap_{i=0}^{k-1} T_i$ precisely represents the set of all executions of M of length k or less, starting with a legal initial state, and $\neg P^k$ is a formula representing the condition that P is violated by a bounded execution of M of length k or less. Hence, φ^k is satisfiable if and only if there exists an execution of M of length k or less that violates the property P: φ^k is typically translated to CNF and solved by a conventional SAT solver.

The formula $\neg P^k$ may be used to express both safety and liveness properties. Liveness properties of the form **AF**p are checked by having $\neg P^k$ represent a loop within a bounded execution of length at most k, such that p is violated on each state in the loop. However, the more common application of BMC is for the purpose of checking safety properties of the form **AG**p (p is some propositional expression). In this case, Eq. 1.1 reduces to $\varphi^k = I \wedge \bigcap_{i=0}^{k-1} T_i \wedge (\bigvee_{i=0}^{k} \neg P_i)$, where P_i is the expression p in time step i. Thus, this formula can be satisfied if and only if for some i ($i \leq k$) there exists a reachable state in time step i in which p is violated. Figure 1.14 shows a circuit representation of this equation, where the block $\bar{P}$ denotes a combinational logic block computing $\neg P^i$ as a function of the state variables of time step i.

Recent research has improved on both the technology and methodology of the basic BMC method in several ways. These techniques attempt to generate a more compact CNF for the BMC problem in the hope that it translates into an easier SAT problem. The *bounded cone of influence* (BCOI) reduction [33] is a variation on the classical *cone of influence (COI)* reduction used in traditional model checking. The intuition is that over a bounded time interval we need not consider every state variable in the classical COI in every time step. Specifically, in Fig. 1.14, the BCOI reduction would extract the transitive fan-in cone of the gate g and construct the BMC-CNF only from this subcircuit. The BCOI reduction is a cheap, easy-to-apply transformation that is often fairly effective in practice.

Another technique uses binary AND–INVERTER graphs to represent the transition relation of the system as well as the unrolled transition relation used for the BMC

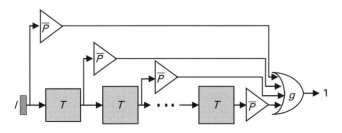

Figure 1.14 Bounded model checking

problem (Fig. 1.14). The graph is compressed as it is built by using an efficient functional hashing scheme across two levels of logic, as well as term rewriting techniques. The CNF for the BMC problem is generated from this compressed representation. The SAT results from earlier BMC runs are used to set appropriate P nodes (Fig. 1.14) to 0 and then rehash the circuit graph to obtain further compression. Such techniques work extremely well in practice, especially if the logic-level circuit used for the verification has been generated through a quick on-the-fly synthesis from an RTL description.

Although BMC is, by its intent, an incomplete bug-finding method rather than a complete verification method, a given property can be certified to be true if no counter-examples are found through BMC, up to the *sequential depth* of the circuit. The sequential depth of a circuit is the length of the *longest* of the shortest paths from the initial state or states to other reachable states of the system. There have been a few attempts at computing or estimating the sequential depth of a circuit, for use as a target depth for BMC. However, the problem of efficiently computing or tightly over-approximating the sequential depth of industrial size arbitrary sequential circuits largely remains an open problem. It is well known that different propositional encodings of the same problem can result in dramatically different run-times on a given SAT solver. The approach of *binary time-frame expansion* proposed by Fallah [34] provides a different propositional encoding of the check for violation of the property in various time-frames of an unrolled circuit.

Several successful attempts at applying SAT-based BMC technology to industrial problems have been reported over the past several years. The original proponents of BMC reported a case study [32] where they applied BMC based on the SAT solvers to verify safety properties on five control units from the PowerPC^{TM} microprocessor. Bounded model checking was found to outperform significantly the BDD-based model checker for several of the benchmarks. Bjesse *et al.* [35] report a significant increase in bug-finding speed and efficiency by their application of SAT-BMC to check safety properties in the memory subsystem of the Alpha microprocessor. A recent comprehensive analysis with respect to the performance and capacity of BMC has also been presented. [36,37]

Please note that the above techniques based on SAT methods can be cast to the corresponding problems for ATPG techniques. Therefore, recursive-learning-based ATPG techniques may efficiently check model designs. In particular, if the design is given as a logic circuit, ATPG-based model checking methods work very well compared with SAT-based ones.

1.4   Equivalence-checking techniques

In this section, I discuss formal equivalence-checking methods for RTL or gate-level designs. As with model-checking, the base methods are based on BDD, SAT, ATPG, or combined methods. I first discuss the definition of equivalence of sequential circuits in VLSI-design processes and then present the equivalence-checking methods for both combinational and sequential circuits.

1.4.1   Definition of equivalent designs

Combinational equivalence checking (CEC) of register-transfer-level (RTL) or gate-level designs is the most widely adopted and successful formal validation technology used in modern-day IC design flows. Register-transfer-level or gate-level circuits arising in the context of IC design flows are usually sequential circuits. It is often the case that two such sequential circuits are compared for equivalence – for example, two copies of the same circuit before and after a sequence of manual or automatic optimization steps, respectively. Several notions of sequential hardware equivalence have been proposed in the literature. However, formal sequential equivalence checking is generally recognized as a fairly intractable problem that cannot be solved efficiently for large industrial designs, except in a few special cases.

Sequential circuits can be represented as finite-state machines (FSMs). An FSM, $F = (I, O, L, S_0, \Delta, \lambda)$, is a 6-tuple, where $I = (x_1, x_2, \ldots, x_m)$ is an ordered set of inputs, $O = (z_1, z_2, \ldots, z_p)$ is an ordered set of outputs, L is an ordered set of state variables (denoting latches), $S_0 \subseteq B^{|L|}$ is a non-empty set of initial states, $\Delta: B^{|L|} \times B^m \to B^{|L|}$ is the next-state function, and $\lambda: B^{|L|} \times B^m \to B^p$ is the output function. A state S of F is a Boolean valuation to the state variables L. In the sequel, the present- and next-state variables corresponding to a latch l will be denoted l and δ_l, respectively. If the two sequential circuits being checked for equivalence share the same set of inputs I, outputs O, and latches L, then it can be shown that it is sufficient to check their *combinational portions* for equivalence. In fact, the two sets of latches do not need to be identical, but there must be some *suitable* mapping between them (this notion is formalized below). Thus, in such a scenario, the sequential equivalence-checking problem can be solved as a sequence of two sub-problems: finding a mapping between the latches of the two circuits, and then checking the combinational portions of the two circuits for equivalence under this mapping. The former is known as the *latch-mapping problem* and the latter as *combinational equivalence checking* (CEC).

1.4.2   Latch-mapping problem

Latch mapping is the first problem to be solved when trying to check sequential equivalence of two circuits using CEC. Informally, the idea is to find a mapping of latches between the two circuits, such that under this mapping (and assuming that the circuits have the same set of input and output signals), the two circuits produce identical output sequences when supplied with the same input sequences. To formalize the discussion, let the two sequential circuits being checked for equivalence be represented by FSMs F_1 and F_2, respectively. Further, to simplify the exposition, we assume that the two circuits have the same identical clock, the same inputs and outputs, and exactly one initial state, denoted $S_{0,1}$ and $S_{0,2}$, respectively. Thus, $F_1 = (I, O, L_1, S_{0,1}, \Delta_1, \lambda_1)$ and $F_2 = (I, O, L_2, S_{0,2}, \Delta_2, \lambda_2)$. Let $L = L_1 \cup L_2$ denote the combined state variables of F_1 and F_2. Further, if S_1 and S_2 are states in the state spaces of F_1 and F_2, respectively – that is, $S_1 \in B^{|L_1|}$ and $S_2 \in B^{|L_2|}$ – we use $S = S_1 \cup S_2$ to denote the combined state. Similarly, the combined transition function Δ is obtained by combining Δ_1 and Δ_2 and the combined initial state $S_0 = S_{0,1} \cup S_{0,2}$.

The latch-mapping problem is posed on the combined set of latch variables L and the combined states in the state-space of these variables. A latch mapping is denoted by a *latch-correspondence relation*, R_L, which is an equivalence relation on the latches, L. Thus, R_L: $L \times L \to B$. Further, the *variable correspondence condition*, V_L: $B^{|L|} \to B$, is a predicate that defines whether a state S conforms to R_L – that is, whether equivalent latch variables assume identical values in S:

$$V_{\mathrm{L}}(S) \Leftrightarrow \forall l_1\, l_2 (R_L(l_1\, l_2) \Rightarrow S(l_1) = S(l_2)).$$

The relation R_L is designed to group together latches that are equivalent, under some notion of sequential equivalence. For the purposes of this exposition, we will use the following definition of R_L, proposed by van Eijk and Jess, [38] based on a sufficient (but not necessary) condition for latch equivalence.

DEFINITION 4.2 (Latch correspondence relation) *[38] A latch correspondence relation is an equivalence relation, R_L: $L \times L \to B$, which satisfies the following conditions:*

- *It is true in the initial state, S_0, of the combined FSM: $V_L(S_0) = 1$;*
- *It is invariant under the next-state function:* $\forall\, S \in B^{|L|},\, X \in B^m$: $R_L(S) \Rightarrow R_{\mathrm{L}}(\Delta(S,X))$.

Methods for latch mapping can be classified as incomplete methods or complete methods. Incomplete methods use heuristics to group promising matches without providing any guarantee on the correctness or completeness of the matching. They can be function-based or non-function-based. Non-function-based incomplete methods (e.g., [39]) use name or structural comparisons to group latches. The rationale for such methods is that combinational optimization, through automatic tools, usually leave net names and much of the combinational structure unchanged. Function-based incomplete methods, such as those proposed in [39] and [40], use random simulation or ATPG-based searches to generate inequivalence information, which is used to group latches. Complete methods, on the other hand, are guaranteed to produce a latch mapping, if one exists, given sufficient computational resources. Almost all complete methods for latch mapping [41] employ a functional fix-point iteration to refine the set of latches into a provably correct and complete grouping. *Van Eijk's algorithm*, [38] shown in Fig. 1.15, is an instance of this class of algorithm. It starts with the set of latch mappings obtained from other methods and tries to increase the set with some methods (heuristics) and then check if the expanded set is valid or not by actually verifying the outputs to the latches. It keeps repeating this process until fix-points are reached.

1.4.3 Practical combinational equivalence checking

Once a latch mapping has been performed on the given pair of FSMs, F_1 and F_2, the next step is to perform combinational equivalence checking on the *combinational portions* of these circuits. Specifically, this involves solving a combinatorial problem

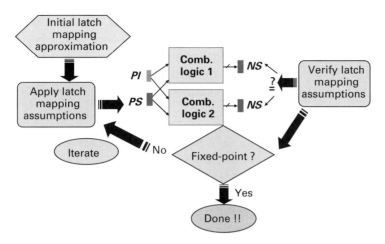

Figure 1.15   Van Eijk's algorithm for latch mapping

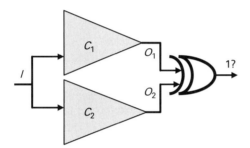

Figure 1.16   Miter construction for combinational equivalence checking

on a circuit called a *miter*, [42] shown in Fig. 1.16, which is constructed as follows. First, the latches in F_1 and F_2 are removed – that is, the sequential feedback loops are cut at the latches. For each latch $l \in L_1 \cup L_2$, the present-state variable l is included in the set of primary input signals and the next-state variable δ_l is included in the set of primary output signals for the respective circuit. Further, each matched set of present-state variables is merged together (i.e., assumed to be driven through a common signal), as for the previously generated latch mapping. Note that we have assumed earlier that the two circuits are driven by the same set of input signals. Hence, in Fig. 1.16, the input signal set I driving the circuits is the set of common primary inputs from the original sequential circuits as well as the set of present-state variable signals from the former latches, merged under the latch mapping. The circuits C_1 and C_2 shown in the figure comprise the combinational logic circuitry implementing the next-state functions Δ and output functions λ of FSMs F_1 and F_2, respectively. The output signal sets O_1 and O_2 comprise the output signals of the respective FSMs, as well as the next-state variables of the former latches. Recall that F_1 and F_2 were assumed to have the same set of outputs and the latch mapping allows a matching of the next-state variables. Thus, in Fig. 1.16, corresponding output signals from O_1 and O_2 are pairwise *exclusive-ORed* and a disjunction of these XOR outputs is taken (denoted by the big XOR gate in the figure). This construction gives us a circuit referred to as a *miter*.

The combinational equivalence-checking problem, then, is to check if there exists an input combination at the signals *I* that causes the miter output to be logic value 1. If not, then the two combinational circuits are equivalent. However, if such an input combination exists, then at least one pair of corresponding outputs in the miter would assume different values under this input. Thus, the two combinational circuits being compared in the miter are not equivalent. Combinational equivalence checking is, theoretically, a co-NP-hard problem and, hence, intractable except for relatively small instances. However, about 20 years ago, researchers working on this problem [43] made the observation that practical instances of this problem are actually more tractable, since the two circuits being checked have a high degree of structural (and hence functional) similarity. This happens because the two circuits are usually different snapshots of the same design picked up from different stages of the design and optimization process. Automatic tools and even manual design steps touch a small portion of the design at a time and frequently preserve the overall logical structure of the design. This single observation revolutionized the scope and usage of combinational equivalence-checking tools in modern RTL design flows.

Almost all industrial CEC tools in use today exploit the notion of structural similarity between the circuits being compared and are based on the principle of *equivalence checking using internal equivalences*. [42,43] The basic idea here is that since the two circuits are structurally fairly similar, there are bound to be internal nodes in the two circuits that functionally correspond with each other. The objective is to detect these internal equivalences and use them to partition the equivalence check on the outputs into a series of smaller and more tractable equivalence checks. To illustrate the principle, let me introduce some notation using the miter in Fig. 1.16 as a basis. Let $I = (i_1, i_2, \ldots, i_n)$ be the common primary inputs of the combinational circuits C_1 and C_2. Let $f_1(i_1, i_2, \ldots, i_n) \in O_1$ and $f_2(i_1, i_2, \ldots, i_n) \in O_2$ be corresponding primary output signals of C_1 and C_2 to be combinationally verified; that is, we would like to check if

$$f_1(i_1, i_2, \ldots, i_n) = f_2(i_1, i_2, \ldots, i_n). \tag{1.2}$$

Let $x_1, x_2, \ldots, x_k$ and $x_1', x_2', \ldots, x_k'$ be corresponding equivalent internal signals in C_1 and C_2, respectively; that is, say we have already verified that

$$x_1(i_1, i_2, \ldots, i_n) = x_1'(i_1, i_2, \ldots, i_n), \tag{1.3}$$

$$x_2(i_1, i_2, \ldots, i_n) = x_2'(i_1, i_2, \ldots, i_n), \tag{1.4}$$

$$\cdots$$

$$x_k(i_1, i_2, \ldots, i_n) = x_k'(i_1, i_2, \ldots, i_n). \tag{1.5}$$

Further, suppose that signals $x_1, x_2, \ldots, x_k$ in C_1 form a cut between the inputs and outputs such that output f_1 can be expressed exclusively in terms of these signals as

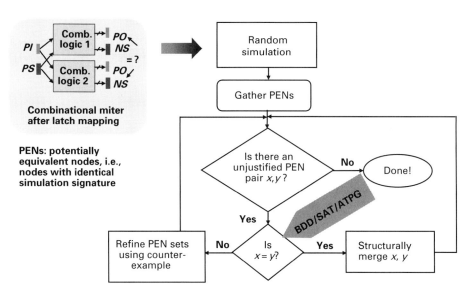

Figure 1.17   General algorithm for CEC using internal equivalences.

$f_1(x_1, x_2, \ldots, x_k)$ and, similarly, f_2 can be expressed as $f_2(x_1', x_2', \ldots, x_k')$. Then, if we can verify that

$$f_1(x_1, x_2, \ldots, x_k) = f_2(x_1, x_2, \ldots, x_k) \quad (k+1)$$

it follows from Eqs. (1.3) to (1.5) that $f_1(i_1, i_2, \ldots, i_n) = f_2(i_1, i_2, \ldots, i_n)$. The rationale of this method is that checking Eq. (1.2), where f_1 and f_2 are expressed monolithically in terms of the entire combinational circuitry of C_1 and C_2, is much more difficult than checking the sequence of equations for x_1 to x_{k+1} which are formulated on much smaller combinational fragments of C_1 and C_2. Thus, given the miter of Fig. 1.16, the overall approach is to proceed topologically from the inputs toward the outputs, identifying internal *potentially equivalent nodes* (PENs) such as x_1 and x_1', x_2 and x_2'; then establish their equivalence (as in Eqs. (1.3)–(1.5)); and then proceed to exploit these to establish the equivalence of topologically deeper PENs (as in the corresponding equation for x_{k+1}, all the way to the primary outputs. Figure 1.17 illustrates this algorithm. Typically, the first step is to perform a quick phase of random simulation on the miter and group together nodes or signals with identical simulation signatures as PENs. These are then validated in topological order. If a pair of PENs is found to be equivalent, these signals (and their input cones of influence) are structurally merged. This reduces the effective size of the miter and increases the efficiency of engines acting on it. If a pair of PENs is found to be inequivalent, the checking engine would typically return an input vector – that is, an assignment to the signals I, under which the two signals assume different values. This is then used to refine the PEN sets by simulating the current miter with this input vector.

Most of the major works in the literature on combinational equivalence checking as well as most commercial offerings in this area today are broadly based on the above algorithm for equivalence checking using internal equivalences. The actual

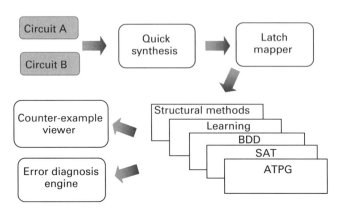

Figure 1.18   A typical modern CEC tool

equivalence checking of each PEN pair is usually performed using one of a variety of engines, including but not limited to BDDs, SAT solvers, and ATPG (automatic-test-pattern-generation-based) structural reasoning, and graph isomorphism checks on the circuit graph. The specific engines used and the heuristics used to guide their orchestration in picking PENs and validating them form the main differences between individual CEC tools. Sometimes these choices can lead to substantial savings in computing resources.

The typical composition of a modern CEC tool is shown in Fig. 1.18. At the core of the tool is a multi-engine solver, comprising, for example, a BDD engine, a SAT solver, an ATPG reasoning engine, a random simulation engine, a host of structural reasoning methods, and a sophisticated set of heuristics for orchestrating these engines to perform the actual equivalence-checking tasks. The input to CEC tools consists of two sequential circuits, one or both of which may be specified at RT level. Since all the engines operate on logic-level circuitry, the typical approach is first to perform a quick synthesis to gate level and then to proceed with equivalence checking of the gate-level circuits. Thus, an RT-gate synthesizer is typically included in the CEC tool, as is a latch mapper to transform the sequential problem to a combinational one. Combinational equivalence checking tools also have comprehensive debugging capabilities to pinpoint error sources when inequivalences are detected, as well as counter-example visualization capabilities, the ability to cross-link RTL and gate-level netlists for easy debugging, and the ability to checkpoint the verification process and restart again from an intermediate checkpoint. By using the PEN-based equivalence-checking method-ology and highly efficient Boolean reasoning engines available today, modern CEC tools can handle circuits of up to a few million gates, flat, in a few hours of run-time.

1.4.4   Sequential equivalence checking (SEC)

If the two sequential circuits to be compared do not have latch mapping, i.e., they have different state encodings, the equivalence problem cannot be reduced to combinational equivalence checking. In such cases, we need to reason about sequential circuits

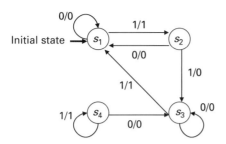

Figure 1.19   An example of a state transition graph of an FSM

directly. Intuitively this is a model-checking problem with the property of saying, "The outputs are always equivalent between the two sequential circuits." Although sequential equivalence checking can be solved by model-checking methods, there are more direct ways to compare the outputs of the two circuits, that is based on forward traversal in model checking. This is called "reachability analysis." Reachability analysis traverses state space in the designs starting from initial states and checks the equivalence of the outputs, either with explicit state traversal or implicit state traversal (symbolic traversal). Usually symbolic state traversal can deal with much larger circuits than explicit state traversal.

First of all, we need to define the sequential equivalence checking problem. Given sequential circuits are first transformed into finite state machines (FSMs) and then FSMs are analyzed. An FSM F is represented as a 6-tuple $<I, S, \delta, S_0, O, \lambda>$, where I represents the set of input signals, S represents the set of states, $\delta: S \times I \rightarrow S$ represents the set of next-state functions, S_0 ($S_0 \subseteq S$) represents the set of initial states, O ($I \cap O = \emptyset$) represents the set of output signals, and $\lambda: S \times I \rightarrow O$ represents the set of functions of output signals. For example, we consider an FSM that consists of:

- $I = \{i_1\}$,
- $S = \{s_1, s_2, s_3, s_4\}$,
- $O = \{o_1\}$,
- $\delta(S \times I) = \{\delta(s_1,0) = s_1, \delta(s_1,1) = s_2, \delta(s_2,0) = s_1, \delta(s_2,1) = s_3,$
  $\delta(s_3,0) = s_3, \delta(s_3,1) = s_1, \delta(s_4,0) = s_3, \delta(s_4,1) = s_4\}$,
- $\lambda(S \times I) = \{\lambda(s_1,0) = 0, \lambda(s_1,1) = 1, \lambda(s_2,0) = 0, \lambda(s_2,1) = 0,$
  $\lambda(s_3,0) = 0, \lambda(s_3,1) = 1, \lambda(s_4,0) = 0, \lambda(s_4,1) = 1\}$,
- $S_0 = \{s_1\}$.

Figure 1.19 shows a state transition graph of an FSM. The reachable states of the FSM can be enumerated explicitly on the state transition graph. Starting from the initial states, reachable states are traversed by considering sequences of input signals. For the example of the FSM shown in Fig. 1.19, we can identify that states s_2 and s_3 are reachable from the initial state s_1 if the input signal i_1 takes the value 1 in states s_1 and s_2. On the other hand, there is no way to reach state s_4 from the initial state, i.e., s_4 is an unreachable state. In general, no unreachable states are dealt with in formal verification.

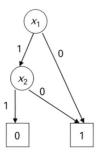

Figure 1.20   BDD representation of the reachability state set

The main drawback of an explicit representation of reachable states is that it cannot represent reachable states of large circuits. For example, it is impossible to make the state transition graph of a sequential circuit that has 100 flip-flops because the circuit may have 2^{100} reachable states. To overcome this problem, most verification techniques use implicit methods, as shown in Section 1.3. In implicit methods, the set of reachable states is represented as a characteristic function that is represented by using BDD or CNF for SAT methods. Therefore, reachable states of large circuits can be represented efficiently. Let us consider the characteristic function of a set of states. For a set of states $S(S = \{s_1, \ldots, s_n\})$ and its subset $S'(S' = \{s_{k1}, \ldots, s_{kn}\}$ $(1 \le ki \le n))$, a function $\chi S' : S \to \{0,1\}$ is defined as:

$$\chi S'(s) = 1 \ (if \ s \in S')$$
$$0 \ (otherwise).$$

The function $\chi S'(s)$ is called the characteristic function of S'. For example, in the FSM of Fig. 1.19, we represent states s_1, s_2, s_3, and s_4 with variables x_1 and x_2 such that $s_1 = \overline{x_1} \cdot \overline{x_2}$, $s_2 = \overline{x_1} \cdot x_2$, $s_3 = x_1 \cdot \overline{x_2}$, and $s_4 = x_1 \cdot x_2$. When the characteristic function of the set of reachable states S' is considered, $\chi S'(s) = 1$ in states s_1, s_2, and s_3 while $\chi S'(s) = 0$ in state s_4. Therefore, the logic function of $\chi S'(s)$ will be $\overline{x_1} + \overline{x_2}$. Figure 1.20 shows the BDD which represents reachable states s_1, s_2, and s_3, where states are encoded with two state variables x_1, and x_2.

Similarly, the characteristic function of the set of next-state functions χ^δ is calculated as follows. When states in S are represented by using k variables $x_1, \ldots, x_k$, the set of next-state functions, $\delta : S \times I \to S$, consists of k next-state functions such that $\delta_i : \{0,1\}^k \times I \to \{0,1\}$. When we represent next-state variables as $x_1', \ldots, x_k'$, the characteristic function $\chi^\delta(x_i', \delta_i)$ is represented as follows:

$$\chi^\delta(x_i', \delta_i) = \prod_{i=1}^{k} (x_i' \equiv \delta_i).$$

Note that $x_i' \equiv \delta_i$ corresponds to $x_i' \cdot \delta_i + x_i' \cdot \delta_i$.

Let me explain how to calculate the characteristic function for the set of next-state functions in the FSM of Figure 1.20. Suppose that next-state functions for next-state

```
BFS_FSM_forward(S, I, δ, S₀) {
    Reached = From = New⁰ = S₀;
    k = 0;δ
    do { δ
        k = k + 1;
        To = Img(δ,From);
        Newᵏ = To – Reached;
        From = Newᵏ;
        Reached = Reached ∪ Newᵏ;
    } while (Newᵏ = 0)
}
```

Figure 1.21   Breadth first-state traversal algorithm for FSM

variables x_1' and x_2' are represented as $\delta_1(x_1, x_2, i_1) = i_1 \cdot x_2 + i_1 \cdot x_1$ and $\delta_2(x_1, x_2, i_1)$ $= i_1 \cdot \overline{x_1} + \overline{x_2}$. The characteristic function of the set of next-state functions is:

$$\chi^\delta(x_1', x_2', \delta_1, \delta_2) = (x_1' \equiv \delta_1)(x_2' \equiv \delta_2)$$
$$\chi^\delta(x1', x2', x1, x2, i1) = \overline{x1'}.\overline{x2'}.\overline{x1}.\overline{i1} + \overline{x1'}.x1.\overline{x2}.i1 + \overline{x1'}.x2^1.\overline{x1}.\overline{x2}.i1$$
$$+ x1'.x1.\overline{x2}.\overline{i1} + x1'.\overline{x2'}.x1.x2.\overline{i1} + x1'.\overline{x2'}.x2.i1.$$

Figure 1.21 shows the algorithm that enumerates reachable states when an FSM is given. The inputs of the algorithm are the set of states S, the set of input signals I, the set of next-state functions δ, and the set of initial states S_0. At the beginning of the algorithm, the set of reached states *Reached*, the set of states that are the source of state transitions *From*, and the set of states traversed after the kth state transitions New^k are initialized by the set of initial states S_0. Then, the following procedures are carried out while $New^k \neq \varnothing$.

1. The set of states *To* that is traversed by one state transition from the states of *From* is calculated. Calculating *To* is called image computation and represented as the function $Img^{(\delta,From)}$. The detail of the function $Img^{(\delta,From)}$ is described later.
2. New^k is calculated by removing the states in *Reached* from the states in *To*.
3. The obtained New^k is set to *From* for the next-state enumeration.
4. Finally, the update of *Reached* by the union of *Reached* and New^k is carried out.

The implementation of the function $Img^{(\delta,From)}$ is different for explicit and implicit methods. In explicit methods, *To* is calculated by enumerating all possible inputs for all states in *From*. On the other hand, in implicit methods, a smoothing operation is carried out to calculate *To*. For a logic function f with n variables ($f(x_1, x_2, \ldots, x_n)$), the smoothing operation $^\exists x_i f$ with respect to variable x_i is defined as follows:

$$^\exists x_i f = f_{x_i} f_{\overline{x_i}},$$

where f_{x_i} is derived by assigning 1 for x_i in function f while $f_{\overline{x_i}}$ is derived by assigning 0. Similarly, a smoothing operation with respect to variables in a set $X = (x_1, \ldots x_n)$ is defined as follows:

$$^\exists X f = ^\exists x_1 (^\exists x_2 (\ldots (^\exists x_n f))).$$

As an example, we apply smoothing operation to variable x_1 in a function $f(x_1, x_2, x_3, x_4) = x_1 \cdot x_2 + \overline{x_1} \cdot x_3 + x_2 \cdot x_3 \cdot x_4 \ldots$

$$f_{x_1} = x_2,$$
$$f_{x_1} = x_3.$$

Therefore,

$$^{\exists}xf(x_1, x_2, x_3, x_4) = x_1 + x_3.$$

The function $Img^{(\delta, From)}$ with smoothing operation is defined as:

$$Img^{(\delta, From)} = ^{\exists} S^{\exists} I(\chi From \cdot \chi \delta).$$

The product of the characteristic functions $\chi From$ and $\chi \delta$ represents the set of states that are traversed from $From$ by one state transition. Therefore, after the application of the smoothing operation to the product with respect to variables in S and I, we can obtain the function that is represented by next-state variables. For example, in the FSM of Fig. 1.19, we calculate the next states traversed from the initial state $\overline{x_1} \cdot \overline{x_2}$ when the next-state functions $\delta^1(x_1, x_2, i_1) = i_1 \cdot x^2 + i_1 \cdot x_1$ and $\delta^1(x_1, x_2, i_1) = i_1 \cdot \overline{x_1} \cdot \overline{x_2}$ are given. The characteristic function for the set of next-state functions χ^δ is represented as:

$$\chi^\delta = (x_1' \equiv \delta_1)(x_2' \equiv \delta_2) = (x_1' \equiv i_1 \cdot x_2 + i_1 \cdot x_1)(x_2' \equiv i_1 \cdot x_1 \cdot x_2).$$

The characteristic function for the initial state χS_0 is represented as:

$$\chi S_0 = \overline{x_1} \cdot \overline{x_2}.$$

Therefore, the next states traversed from the initial state $\overline{x_1} \cdot \overline{x_2}$ are calculated by the product of χ^δ and χS_0.

$$\chi^\delta \cdot \chi S_0 = (x_1' \equiv i_1 \cdot x_2 + \overline{i_1} \cdot x_1)(x_2' \equiv i_1 \cdot \overline{x_1} \cdot \overline{x_2}) \cdot (\overline{x_1} \cdot \overline{x_2})$$
$$= (x_1' \equiv 0)(x_2' \equiv i_1 \cdot \overline{x_1} \cdot \overline{x_2})$$

Then, the smoothing operation with respect to the variables in sets S and I is applied.

$$^{\exists}I\chi\delta \cdot \chi S_0 = (x_1' \equiv 0)(x_2' \equiv x_1.x_2)$$
$$^{\exists}S^{\exists}I\chi\delta \cdot \chi S_0 = \overline{x_1'}.x_2' + \overline{x_1'} \cdot \overline{x_2'}.$$

As a result, we can identify that the next states for s_1 are states $s_1(\overline{x_1'} \cdot \overline{x_2'})$ and $s_2(\overline{x_1'} \cdot \overline{x_2'})$.

Equivalence checking of two sequential circuits verifies whether the behavior of two given FSMs $M_1 = <I,S,\delta_1,S_0,O,\lambda_1>$ and $M_2 = <I,T,\delta_2,T_0,O,\lambda_2>$ is equivalent or not. This is equivalent to checking whether there exists an input signal that leads to a different output signal when the same input sequence is given from the initial states of M_1 and M_2. Note that both M_1 and M_2 have the same set of input signals I and the same set of output signals O. This problem is considered by using the product machine of M_1 and M_2 shown in Fig. 1.22.

In the product machine, the XNOR of output signals is calculated for all input sequences. The output signals of M_1 and M_2 are equivalent when the XNOR of each pair of output signals is 1. On the other hand, M_1 and M_2 are not equivalent when there

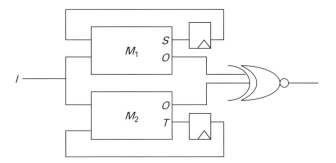

Figure 1.22   Equivalence checking on two sequential circuits

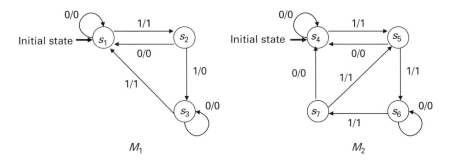

Figure 1.23   FSMs extracted from the circuits in Fig. 1.22

exists a pair of output signals such that XNOR is 0. Therefore, we check whether a pair of states in M_1 and M_2 such that XNOR is 0 is reachable from the initial states of the product machine or not.

The product machine of M_1 and M_2, $M_{12} = <I,U,\delta_{12},U_0,\{0,1\},\lambda_{12}>$ is defined as:

- I: *the set of inputs*,
- $U = S \times T$,
- δ_{12}: $(S \times T) \times I \rightarrow (S \times T) = (\delta_1\colon S \times I \rightarrow S,\ \delta_2\colon T \times I \rightarrow T)$,
- $U_0 = S_0 \times T_0$,
- $\{0,1\}$: the *output value of XNOR*,
- λ_{12}: $(S \times T) \times I \rightarrow \{0,1\}$ (1 *if* $\lambda_1(S,I) = \lambda_2(T,I)$, 0 *otherwise*.

Let us verify the equivalence of the two FSMs shown in Fig. 1.23. Note that the set of reachable states on the product machine is calculated based on the implicit method. States of FSM M_1 are represented as $s_1 = \overline{x_1} \cdot \overline{x_2}$, $s_2 = \overline{x_1} \cdot x_2$, and $s_3 = x_1 \cdot \overline{x_2}$. When we represent the input signal as i_1, the next-state functions δ_1 and δ_2 ($\delta_1 = \{\delta_1, \delta_2\}$) and the output function $\lambda_1(\lambda_1 = \{\lambda_1\})$ are represented as:

$$\delta_1 = i_1.x_2 + \overline{i_1}.x_1,$$
$$\delta_2 = i_1.\overline{x_1}.\overline{x_2},$$
$$\lambda_1 = i_1.\overline{x_2}.$$

On the other hand, states of FSM M_2 are represented as $s_4 = \overline{x_3} \cdot \overline{x_4}$, $s_5 = \overline{x_3} \cdot x_4$, $s_6 = x_3 \cdot \overline{x_4}$, and $s_7 = x_3 \cdot x_4$. The next-state functions δ_3 and δ_4 ($\delta_2 = \{\delta_3, \delta_4\}$) and the output function $\lambda_2(\lambda_2 = \{\lambda_2\})$ are represented as:

$$\delta_3 = x_3.\overline{x_4} + i_1.\overline{x_3}.x_4,$$
$$\delta_4 = i_1.x_3 + i_1.\overline{x_4},$$
$$\lambda_2 = i_1.$$

The output function of the product machine is calculated as:

$$\overline{\lambda 1 \oplus \lambda 2} = \overline{(i_1.\overline{x_2})} \, \overline{i_1} + i_1.\overline{x_2}$$
$$= (\overline{i_1} + x_2)\overline{i_1} + i_1.\overline{x_2}$$
$$= \overline{i_1} + \overline{x_2}.$$

The function implies that the value of XNOR is 1 when $i_1 = 0$, regardless of the state variables. On the other hand, the value of XNOR depends on the value of $\overline{x}_2$ when $i_1 = 1$. The outputs of M_1 and M_2 are different in the states where $x_2 = 1$ because in such states XNOR will be 0. Therefore, we check whether those states are reachable from the initial states of M_{12} or not.

Suppose that the next-state variables of M_1 and M_2 are represented as $x_1{}'$, $x_2{}'$, $x_3{}'$, and $x_4{}'$. The characteristic function of the set of next-state functions for M_{12} is represented as:

$$\chi \delta_{12} = \chi \delta_1.\chi \delta_2 = (x_1' \equiv \delta_1)(x_2' \equiv \delta_2)(x_3' \equiv \delta_3)(x_4' \equiv \delta_4)$$
$$= (x_1' \equiv i_1.x_2 + \overline{i_1}.x_1)(x_2' \equiv i_1.\overline{x_1}.\overline{x_2})$$
$$(x_3' \equiv x_3.\overline{x_4} + i_1.\overline{x_3}.x_4)(x_4' \equiv i_1.x_3 + i_1.\overline{x_4}).$$

The characteristic function of the initial state s_1:s_4 is represented as:

$$\chi U_0 = \chi S_0.\chi T_0 = \overline{x_1}.\overline{x_2}.\overline{x_3}.\overline{x_4}.$$

Therefore, the set of the next states traversed from the initial state s_1:s_4 is calculated as:

$$\chi \delta_{12}.\chi U_0 = (x_1' \equiv i_1.x_2 + \overline{i_1}.x_1)(x_2' \equiv i_1.\overline{x_1}.\overline{x_2})(x_3' \equiv x_3.\overline{x_4} + i_1.\overline{x_3}.x_4)$$
$$(x_4' \equiv i_1.x_3 + i_1.\overline{x_4})(\overline{x_1}.\overline{x_2}.\overline{x_3}.\overline{x_4})$$
$$= (x_3' \equiv 0)(x_2' \equiv i_1.\overline{x_1}.\overline{x_2})(x_3' \equiv 0)(x_4' \equiv i_1.\overline{x_4}).$$

Then, we apply the smoothing operation to the variables in sets U and I.

$$^{\exists}I\chi \delta_{12}.\chi U_0 = (x_1' \equiv 0)(x_2' \equiv \overline{x_1} . \overline{x_2})(x_3' \equiv 0)(x_4' \equiv \overline{x_4}),$$
$$^{\exists}U^{\exists}I\chi \delta_{12}.\delta U_0 = \overline{x_1'}.\overline{x_2'}.\overline{x_3'}.\overline{x_4'} + \overline{x_1'}.x_2'.\overline{x_3'}.x_4'.$$

As a result, we can identify that the next states of the initial state s_1:s_4 are states s_1:s_4 and s_2:s_5. Since $x_2{}'$ is 1 in state s_2:s_5, the product machine produces 0. This means that the two FSMs are inequivalent. The state transition graph of the product machine is shown in Fig. 1.24.

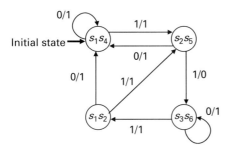

Figure 1.24   Product FSM generated from the two FSMs in Fig. 1.23

1.5   Techniques for higher-level design descriptions

In this section, we introduce equivalence-checking methods for design descriptions that are higher level than RTL. High-level design descriptions are represented in program-like formats, and design languages based on C/C++ languages are typically used. [44,45] To reason about C/C++ language constructs, not only Boolean reasoning but also so-called word-level reasoning is essential for efficient verification. For example, one integer variable is a single variable, although it must be expanded into 32 Boolean variables if Boolean reasoning is applied. If we always expand such variables into multiples of Boolean variables, the number of variables for Boolean reasoning, like the ones based on SAT solvers and BDD-based routines, easily becomes too large to be processed. Instead, any reasoning procedures on high-level design descriptions should apply word-level analysis methods, which deal as much as possible with all word-level variables as they are. If they somehow fail, analysis methods are switched to Boolean-based ones.

There are decision procedures, such as CVC [46], that can deal with word-level variables. Although they may be based on Boolean SAT solvers as their final reasoning engines, they try to use word-level analysis as much as possible. Here, I concentrate on the use of such decision procedures on equivalence checking for high-level design descriptions.

Another important issue in high-level equivalence checking is the fact that the two design descriptions being compared are typically very similar, since the design processes in high levels consist of a series of small design refinements. If equivalence checking is applied to the descriptions before and after each such small refinement, the difference between the two design descriptions is very small, in the sense that most of the descriptions are the same and there are many internal equivalent corresponding variables. This is basically the same situation as the equivalence checking on two combinational circuits, discussed earlier, which is widely used for formal verification nowadays in industry. Therefore, by partitioning the given design descriptions into much smaller ones through the equivalent variables, the equivalence-checking problem becomes a collection of many small ones. This gives us the ability to deal with the large and practical design descriptions used in industry.

The basic method used to compare two high-level design descriptions is *symbolic simulation*. Since word-level analysis methods should be used as much as possible, symbolic simulation – where each variable is given symbolic values instead of concrete values – can easily accommodate word-level reasoning procedures, such as decision procedures. Also, if necessary, Boolean reasoning can also be incorporated into symbolic simulation in the same way as word-level reasoning.

Here, I briefly review the high-level design flow from the viewpoint of equivalence-checking technology. Then I present symbolic simulation for high-level design descriptions, followed by an introduction of a couple of improved equivalence-checking algorithms based on symbolic simulation that utilize the similarity of the two descriptions to be compared.

Verification of designs is one of the most important tasks in the design of large and complicated systems. Target designs are becoming larger and more complex as integration technologies rapidly improve. This trend makes the verification of the whole design more and more difficult – so much so that design times are dominated by their verification times. Therefore, it is very important to try to verify design descriptions at as high a level as possible. In general, the higher the level of the design description, the smaller the number of components to be analyzed when they are verified. When a description of a design is changed for some reason, it is possible that an error has been introduced into the design. If such an error is found in the later stages of the design flow, design productivity is decreased significantly, because the modification that may be required at the higher-level descriptions may entail going back to the initial stages of the design process. To solve this problem, the error should be sought and corrected as early as possible before implementation. This implies that formal equivalence checking of design descriptions before and after transformations of design descriptions is one of the most important issues in higher-level design stages.

I now present formal equivalence-checking methods for two C descriptions. The basic verification engine for equivalence checking of high-level design descriptions is symbolic simulation. Given two C descriptions, symbolic simulation-based methods verify whether variables corresponding to output signals in a design are equivalent or not, when all variables corresponding to input signals are assumed to be equivalent. As a result of symbolic simulation, variables that are identified as equivalent to each other in the two descriptions are collected into the same equivalence class. Therefore, we can prove the equivalence of variables corresponding to output signals by checking whether they are in the same equivalence class or not.

In general, formal methods, including symbolic simulation, will fail when dealing with very large designs. To solve this problem, in the method discussed here, textual differences between descriptions are utilized to reduce the number of equivalence checks of variables. This means that only the variables related to textual differences are verified during symbolic simulation. Therefore, this method is particularly efficient when the two descriptions are similar to each other, because we can expect that there will be few equivalence checks carried out during symbolic simulation. As noted earlier, this is essentially the same strategy used in combinational equivalence-checking methods now commonly used in industry. Equivalence checking on descriptions of large designs is essentially like partitioning large descriptions into a collection of much smaller ones. Therefore, in general, the more similar the two descriptions to be compared, the more efficient the equivalence-checking processes.

Symbolic simulation has become one of the most common techniques in hardware verification. Since variables in the descriptions are treated as symbols rather than as concrete-valued bit vectors, symbolic simulation can efficiently verify larger descriptions than traditional logic simulation. Here, I present a symbolic simulator for the equivalence checking of two C descriptions. The characteristics of the extended symbolic simulator are as follows [47]:

1. Symbolic simulation starts from the beginning of the descriptions.
2. When an expression is simulated symbolically, an equivalence class *(EqvClass)* for the expression is created.
3. If two variables in different EqvClasses are proved to be equivalent during symbolic simulation, the two EqvClasses are merged into a single EqvClass.
4. When a case split occurs owing to conditional statements in the C descriptions, all potentially executable paths are simulated.
5. Functions can be uninterpreted in symbolic simulation. Two uninterpreted function calls to the same function are assumed to be equivalent when all their arguments are equivalent. This is everything we assume on uninterpreted functions. If necessary, interpretation can be introduced to such functions so that more detailed reasoning can be made.
6. After symbolic simulation, the two variables are equivalent if they belong to the same EqvClass.

A simple example of equivalence checking in terms of symbolic simulation is shown in Fig. 1.25. In this example, we can verify the equivalence of the variable reg_0 in the two given descriptions. Initially, the variables reg_1 and reg_2 are assumed to be equivalent in

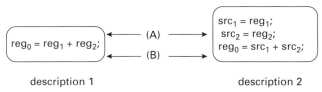

Assumption: the variables reg1 and reg2
are equivalent in both descriptions.

$reg_0 = reg_1 + reg_2;$ ←— (A) —→ $src_1 = reg_1;$
←— (B) —→ $src_2 = reg_2;$
$reg_0 = src_1 + src_2;$

description 1   description 2

Transitions of EqvClasses

Beginning of simulation   (A)
(from assumption) $E_1 = (reg_{1\_1}, reg_{1\_2}, src_{1\_2})$
$E_1 = (reg_{1\_1}, reg_{1\_2})$ —→ $E_2 = (reg_{2\_1}, reg_{2\_2}, src_{2\_2})$
$E_1 = (reg_{2\_1}, reg_{2\_2})$

End of simulation   (B)
$E_1 = (reg_{1\_1}, reg_{1\_2}, src_{1\_2})$ $E_1 = (reg_{1\_1}, reg_{1\_2}, src_{1\_2})$
$E_2 = (reg_{2\_1}, reg_{2\_2}, src_{2\_2})$ ←— $E_2 = (reg_{2\_1}, reg_{2\_2}, src_{2\_2})$
$E'_3 = (reg_{0\_1}, reg_{0\_2}, reg_{1\_1} + reg_{2\_1})$ $E'_3 = (reg_{0\_1}, reg_{1\_1} + reg_{2\_1})$
                                           $E_4 = (reg_{0\_2}, src_{1\_2} + src_{2\_2})$

reg_{0_1} and reg_{0_2} are in the same EqvClass

Figure 1.25   Example of equivalence checking based on symbolic simulation

both descriptions, because these variables correspond to input signals. These assumptions are expressed in the two EqvClasses, E_1 and E_2. Note that we denote a variable v in description 1 as v_1 and in description 2 as v_2. At first, expressions for the variables src_1 and src_2 in description 2 are simulated before reaching point (A). This results in src_{1_2} being inserted into E_1 and src_{2_2} into E_2, because src_{1_2} is equal to reg_{1_2}, and src_{2_2} is equal to reg_{2_2}. Then, two additional EqvClasses, E_3 and E_4, are created before reaching point (B). Finally, reg_{1_1} and reg_{2_1} are substituted for src_{1_2} and src_{2_2} in E_4, respectively, because from E_1 and E_2 we find out that src_{1_2} is equivalent to reg_{1_1}, and src_{2_2} is equivalent to reg_{2_1}. This means that E_3 and E_4 are equivalent to each other. Therefore, E_3 and E_4 are merged into a new EqvClass, E'_3. As a result, we can conclude that the variable reg_0 is equivalent in both descriptions, because the occurrences of reg_0 in both descriptions are in the same EqvClass.

In simple symbolic simulations, the equivalence of the following pairs of expressions cannot be directly proved, because symbolic simulation does not interpret the functionality of the expressions.

$$a + a, 2 * a,$$
$$(a + b) + c, a + (b + c),$$
$$a * (b + c), a * b + a * c.$$

To prove the equivalence of these expressions, the method calls for some sort of decision procedure, such as a cooperating validity checker (CVC). [46] This is a decision procedure that checks the logical validity of given formulas. Formulas are represented by propositional operators and equations between linear mathematical expressions. Such decision procedures can accept quantifier-free formulas in first-order logic. In addition, the formulas can have the following:

- Linear real arithmetic formulas: the supported operators are addition, subtraction, multiplication by a constant, division by a constant, equality, and inequality,
- Real arrays,
- Inductive data types (for example, lists and trees).

We can improve the ability of equivalence checking between variables by using decision procedures in the symbolic simulation for analysis of the simulation results. Compared with substitution used in symbolic simulation, decision procedures generally take longer to compute equivalence because they utilize several theorems to check validity.

To narrow the areas for symbolic simulation, program slicing [48] can be used as preprocessing. It is an operation that identifies semantically meaningful decompositions of programs. In symbolic simulations, program slicing can be used to extract all expressions that are relevant to the difference between the two descriptions to be compared. As a result, the equivalence checking of two descriptions is reduced to the verification of the extracted variables. Program slicing can be used in the context of symbolic simulation in the following ways. Backward slicing for a variable v extracts all expressions that affect the variable v. Forward slicing for a variable v, on the other hand, extracts all expressions that are affected by the variable v. Chopping from a variable s to a variable t is the product set of the forward slice for s and the backward

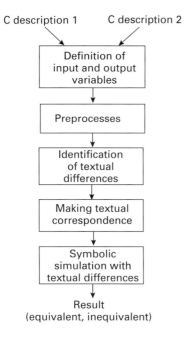

Figure 1.26   The flow of equivalence checking based on textual difference

slice for t. In symbolic simulations, chopping is initially applied to each description from input variables to output variables. Therefore, all expressions relevant to variables for input and output signals in the descriptions are extracted by chopping. As a result, we can avoid wasteful verification of statements that are irrelevant to the variables of input and output signals. In addition to the chopping operation, computing successors, some sorts of forward slicing, can be carried out so that successors for a variable v are all expressions that are directly affected by v.

The flow of equivalence checking based on identification of textual differences in the two design descriptions [49] is shown in Fig. 1.26. As initial inputs, two designs to be compared are given as functions written in C. The variables corresponding to input and output signals in the functions (called *input variables* and *output variables*, respectively) are defined by designers. The methods verify whether all output variables are equivalent when all input variables are assumed to be equivalent.

After input variables and output variables are given, chopping is carried out from input variables to output variables. This operation extracts only parts of descriptions that are affected by input variables and that affect output variables. Therefore, only the extracted descriptions are verified during symbolic simulation.

There are restrictions on the descriptions in C/C++ languages that can be verified, as the targets are hardware design descriptions. These restrictions make the equivalence-checking problems considerably easier and able to deal with realistic sizes of designs. The designs to be verified are allowed to have the following elements:

- All operators (they are not interpreted in symbolic simulation),
- Arrays,

- Assignments including compound assignments,
- *If-then-else* conditional branches,
- Functions and function calls,
- *For* loops and *while* loops (they are unrolled before symbolic simulation),
- No pointer uses (or all pointer uses are analyzed and replaced by certain variables),
- No dynamic memory allocation,
- No recursive function calls.

Though a symbolic simulator can receive all types of operator, subsets of operators can be understood by the decision procedures that are used to decide the equivalence classes. If, however, the decision procedures cannot understand an operator in a formula, they may return the result that the descriptions being compared are not equivalent (fail to show the equivalence). In such cases, the method may return with false-negative results. The method verifies whether or not the behaviors of the given descriptions are equivalent. Therefore, the data types of variables and the problems of overflow or underflow cannot be checked in this method.

In addition, we assume that the given descriptions have the same control flows, with the same correspondence between them, as explained in the following. This is because we assume that the design flow is a collection of small design refinement steps and that the given descriptions have only few differences. First of all, for convenience, several preprocesses, such as in-lining of macro definitions, are carried out on the given descriptions. This can be done by C compilers' preprocessors with the appropriate options. Then, the user-defined functions that do not affect functionalities of designs are removed from the descriptions. For example, input or output functions such as *scanf* and *printf* are removed. When there are loop structures in the descriptions, these must be unrolled using the symbolic simulation methods shown in this chapter. If the number of iterations of a loop is fixed, the loop is unrolled the same number of times as the number of iterations. On the other hand, if the number of iterations is infinite or dependent on input variables, the number of unrollings is specified by users. The equivalence checking will be performed up to this number of iterations for the loop descriptions. If the number of unrollings is not large enough, some possible execution paths in the original descriptions may not exist in the descriptions after loop unrolling. Therefore, the completeness of the equivalence checking depends on the number of unrollings, if loop unrolling is carried out.

After the preprocesses, textual differences between the two given descriptions are identified. This can be done in many ways. The simplest way is to use the standard UNIX command *diff*, which is what I have done here. After textual differences are identified, we can take textual correspondence between descriptions. By using information on textual differences, we can establish a one-to-one correspondence between expressions in the two descriptions. This is based on the assumption that the two design descriptions are not much different. If they are, the one-to-one mapping generation may simply fail, which is not dealt with here. Figure 1.27 shows an example of the textual correspondence between the descriptions. If the corresponding expressions are textually equivalent, they are marked as "*E*." If the corresponding expressions are

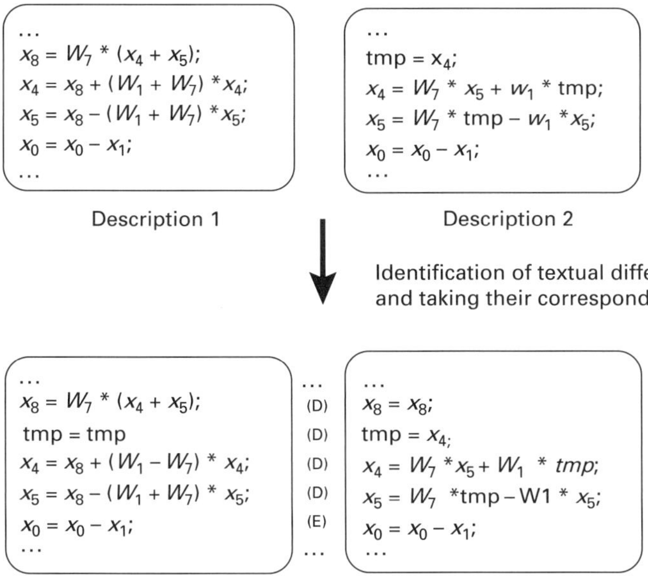

Description 1

$$x_8 = W_7 * (x_4 + x_5);$$
$$x_4 = x_8 + (W_1 + W_7) * x_4;$$
$$x_5 = x_8 - (W_1 + W_7) * x_5;$$
$$x_0 = x_0 - x_1;$$

Description 2

$$tmp = x_4;$$
$$x_4 = W_7 * x_5 + w_1 * tmp;$$
$$x_5 = W_7 * tmp - w_1 * x_5;$$
$$x_0 = x_0 - x_1;$$

Identification of textual difference
and taking their correspondence

$$x_8 = W_7 * (x_4 + x_5);$$ (D) $$x_8 = x_8;$$
$$tmp = tmp$$ (D) $$tmp = x_4;$$
$$x_4 = x_8 + (W_1 - W_7) * x_4;$$ (D) $$x_4 = W_7 * x_5 + W_1 * tmp;$$
$$x_5 = x_8 - (W_1 + W_7) * x_5;$$ (D) $$x_5 = W_7 * tmp - W1 * x_5;$$
$$x_0 = x_0 - x_1;$$ (E) $$x_0 = x_0 - x_1;$$

Figure 1.27   Example of correspondence between expressions in the descriptions

textually different, they are marked as "*D*." Like the expression for the variable *tmp* in description 2 of Fig. 1.27, if an assignment appears in only one of the descriptions, a dummy assignment, such as

$$tmp = tmp,$$

is inserted in the other description to create the correspondence. With this matching process, the two descriptions will have the same number of statements.

To ensure textual correspondence between descriptions, my proposed method will only handle two descriptions that have the same control flows. In other words, we can verify the equivalence of a refinement carried out on a design, as long as it does not change the control flow of the design. If there are small differences in control flow, another type of matching process may be applied before symbolic simulation. If the difference is large, however, my proposed method does not work.

After the processes described above are completed, symbolic simulation to check the equivalence of output variables is carried out. Earlier, I introduced equivalence checking in terms of symbolic simulation. To find equivalent variables, every EqvClass is checked whenever a new EqvClass is created. This means that equivalence checking of variables increases with the square of the size of simulated descriptions. To reduce the number of equivalence checks of variables between the descriptions, my proposed method uses textual differences, which are identified before simulation.

The flow of the algorithm to check the equivalence of a pair of expressions is shown in Fig. 1.28. Depending on whether the pair is marked "E" or "D," the way to simulate and create the EqvClass is different. If the pair is marked "E" and is not affected by variables whose equivalence is not proved, a new EqvClass for the pair is created without checking

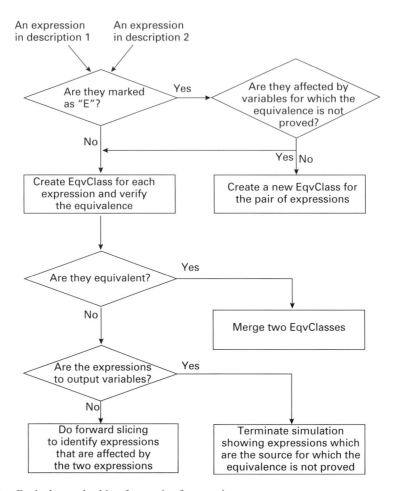

Figure 1.28   Equivalence checking for a pair of expressions

the equivalence. If the pair is marked "D" or is affected by variables whose equivalence is not proved, the equivalence between expressions is verified. After the verification, if the expressions are proved to be equivalent, the two EqvClasses for the expressions are merged. Otherwise, my proposed method evaluates whether these expressions are for output variables or not. If these expressions are assignments for output variables, my method terminates verification and shows all EqvClasses created during symbolic simulation as a counter-example. If, however, the expressions are assignments not for output variables, successors for the pair of simulated expressions are computed by using program slicing in order to identify expressions that are directly affected by this pair. Later, when the simulation reaches the expressions identified as successors for non-equivalent variables, the equivalence of variables assigned by these expressions must be verified, because such variables are affected by the variables whose equivalence is not proved.

   In the equivalence-checking method, equivalence checking of variables is omitted if pairs of expressions are textually equivalent and not affected by variables whose

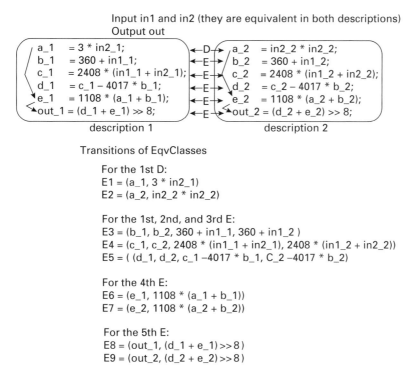

Input in1 and in2 (they are equivalent in both descriptions)
Output out

description 1

a_1	= 3 * in2_1;
b_1	= 360 + in1_1;
c_1	= 2408 * (in1_1 + in2_1);
d_1	= c_1 − 4017 * b_1;
e_1	= 1108 * (a_1 + b_1);
out_1	= (d_1 + e_1) >> 8;

description 2

a_2	= in2_2 * in2_2;
b_2	= 360 + in1_2;
c_2	= 2408 * (in1_2 + in2_2);
d_2	= c_2 − 4017 * b_2;
e_2	= 1108 * (a_2 + b_2);
out_2	= (d_2 + e_2) >> 8;

Transitions of EqvClasses

For the 1st D:
E1 = (a_1, 3 * in2_1)
E2 = (a_2, in2_2 * in2_2)

For the 1st, 2nd, and 3rd E:
E3 = (b_1, b_2, 360 + in1_1, 360 + in1_2)
E4 = (c_1, c_2, 2408 * (in1_1 + in2_1), 2408 * (in1_2 + in2_2))
E5 = ((d_1, d_2, c_1 −4017 * b_1, C_2 −4017 * b_2)

For the 4th E:
E6 = (e_1, 1108 * (a_1 + b_1))
E7 = (e_2, 1108 * (a_2 + b_2))

For the 5th E:
E8 = (out_1, (d_1 + e_1) >> 8)
E9 = (out_2, (d_2 + e_2) >> 8)

Figure 1.29   A simple equivalence-checking example

equivalence is not proved. Therefore, the present method is very efficient when two given descriptions are close to each other, because the equivalence checking between variables is applied only a few times. As a result, we can reduce the verification time significantly. I explain the present method with a simple example, shown in Fig. 1.29. Initially, the input variables in_1 and in_2 are assumed to be equivalent in both descriptions. We verify whether the output variable *out* is equivalent (or not) in both descriptions. Note that after textual correspondence is taken, all variables in description 1 are denoted as *v_*1, whereas all variables in description 2 are denoted as *v_*2.

In the first "D," two EqvClasses for *a_1* and *a_2* are created. Then, the equivalence of *a_1* and *a_2* is verified. Since they are not equivalent, successors for *a_1* and *a_2* are computed to identify expressions that are directly affected by *a_1* and *a_2*. The assignments to the variable *e_1* are identified in description 1, whereas the assignments for the variable *e_2* are identified in description 2. In the first, second, and third "E," three EqvClasses are created without checking the equivalence of *b_1* and *b_2*, *c_1* and *c_2*, and *d_1* and *d_2*. This is because corresponding expressions are textually equivalent, and they are not affected by variables whose equivalence is not proved. In the fourth "E," two EqvClasses for the variables *e_1* and *e_2* are created separately, although they are marked "E." This is because these variables are affected by nonequivalent variables *a_1* and *a_2*. Then, we can identify that the variables *e_1* and *e_2* are not equivalent by equivalence checking. Therefore, successors for *e_1* and *e_2* are computed. As a result, the assignments to the variables *out_1* and *out_2* are identified. Finally, in the last "E," two EqvClasses for variables *out_1* and *out_2* are created. Since they are not equivalent because of the effect from *e_1* and *e_2*, we can conclude that the output variable *out* is not equivalent between descriptions.

So far, I have presented equivalence-checking methods for two C descriptions by means of symbolic simulation. To verify the equivalence efficiently, the method identifies textual differences between two descriptions and utilizes them well so that the number of equivalence checks can be drastically reduced. The method is particularly useful when two large descriptions with few differences are given. The method, however, still traverses all statements from the beginning to the end – although textual differences are used to skip statements with no change. To obtain more efficient equivalence checking, it is necessary to start from each difference (such as a textually different statement) to prove the equivalence, instead of traversing all statements. If the differences are proved to be equivalent, then no further analysis is needed. If some of the differences are not proved to be equivalent, the area to be analyzed may have to be extended so that equivalence can be proved in the extended areas. This extended process can continue until the equivalence is proved or the extension reaches the primary inputs or outputs. In the latter cases, non-equivalence has been proved.

This extension-based method could be much more efficient in cases where large design descriptions have only small differences and they are equivalent. If they are not equivalent, that is the worst case for this method in general, since we have to continue extension until we reach primary inputs or outputs. The overall flow of the extension-based equivalence-checking method is shown in Fig. 1.30 [50]. As inputs, two C programs are given, with the definition of input and output variables. In addition, the correspondence of those variables between programs is given. Then, my method verifies the equivalence of the output variables by using symbolic simulation and reports the verification result ("equivalent" or "not equivalent"). Textual difference identification can be performed in the same way as above – for example, with the use of the UNIX *diff* command. Also, for the purpose of creating correspondence between statements in both descriptions, dummy statements are inserted into the descriptions in the following cases:

- When an assignment is removed, the assignment to the same variable, such as *a = a;* is inserted.
- When a conditional branch is removed, the same branch structure is inserted where all assignments are replaced by ones to the same variable.

Since these inserted statements clearly preserve the original behavior, the result of verification is not changed. Even if many statements are different, the descriptions after the inserted dummy statements cannot be twice as large as the original descriptions.

Then, program-slicing techniques are applied, and the verification area is extracted. The initial verification area for a difference is two sets of statements corresponding to the difference (one set from each description). Note that a difference may consist of

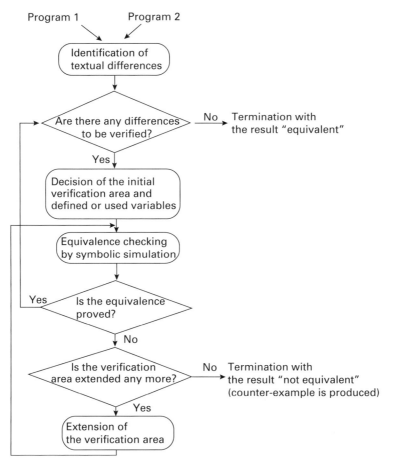

Figure 1.30   The extension-based equivalence-checking algorithm

several statements. We can define input variables and output variables of a local verification area as follows:

- *Local input variable* A variable corresponding to a data-dependence edge coming from outside the verification area and into the area,
- *Local output variable* A variable corresponding to a data-dependence edge coming from inside the verification area and out to the area.

Only when a variable is a local output variable in each description is its equivalence checked in the verification. Although other local output variables are not checked for this difference, they will be taken into account in verification for other differences, if required. A pair of corresponding local input variables is equivalent in the following cases:

- They are not affected by any differences that are proved to be non-equivalent.
- They are already proved to be equivalent by the verification of another difference.

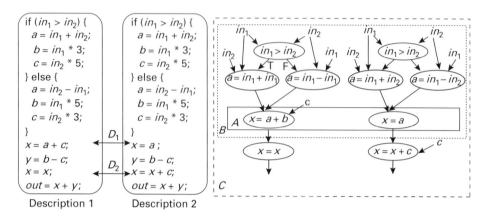

Description 1   Description 2

Figure 1.31   Example of equivalence checking based on extension of verification areas

In the verification, equivalences of other pairs of local input variables are considered to be unknown variables. If all pairs of local output variables are proved to be equivalent, the verification area of the difference is also proved to be equivalent. On the other hand, if the equivalence of any local output variables is not proved, the verification area is extended so that preceding or succeeding statements are included. If the equivalence for a local verification area is not proved, the area is extended based on the dependence relation given by program-slicing techniques. The extension is required because the equivalence of a difference can be proved after extending the verification area.

There are three types of extension for the verification areas: backward extension, forward extension along data dependence, and forward extension along control dependence. In extension, multiple statements that present assignments to the same variable are added to the verification area when their control dependences are different. In such cases, the nodes that control these assignments are also added. After the extensions, the local input and output variables are derived for the new verification area, and verification is carried out. I now show how the extension-based method works using an example shown in Fig. 1.31. We assume that the variables in_1 and in_2 are the primary inputs of the program, and the variable out is the primary output. The statement $x = x$; in description 1 is added as a dummy statement to make a correspondence to $x = x + c$; in description 2.

First, the first difference D_1 is verified. The first verification area in the figure is A; its local input variables are a and c, and its local output variable is the variable x. Since all local input variables are unknown, the equivalence of x cannot be proved. Thus, in this case, we decide to extend the area backward from a. Then, the extended verification area becomes the area B, and the verification is carried out again. In this case, the local input variables are in_1, in_2, and c, and the local output variables are x and $(in_1 > in_2)$. Since the equivalence of x cannot be proved after the verification with the area B, we decide to extend the area forward from x and obtain the area C. After the verification with this area C, we can prove the equivalence of x. The verification for the difference D_2 is not carried out, since it is included in the verification for D_1. Then, as the difference is all verified, it can be said that the two descriptions are functionally equivalent.

In general, a verification area can have multiple local input and output variables. Therefore, there are a number of different ways to apply backward and forward extensions. This makes it difficult to define the best strategy for extensions. In the following, I list some reasonable strategies for extensions that commonly occur in practice:

- Apply backward extensions until the start points of the programs, and then apply forward extensions until the end points,
- Apply forward extensions and backward extensions in turn,
- First apply backward extensions m times, and then apply forward extensions n times (m and n are predefined numbers).

These strategies are similar to ones in equivalence checking of gate-level circuits.

In practical cases, designers know which kinds of refinement are carried out. In such cases, a specific strategy for the refinement can be applied to improve the verification speed. By incorporating this type of knowledge into the equivalence-checking techniques, highly efficient comparisons can be made for high-level design descriptions, which potentially gives dramatic reductions of design bugs found in the later design stages. It is always most important to eliminate as many design errors (bugs) as possible in as early design stages as possible. In this sense, formal verification in higher-level design stages can take the most important role in the total VLSI design flow. This C/C++-based design methodology, which consists of many small refinements of designs, is now emerging and the need of its formal verification support is becoming indispensable [51].

1.6   References

[1] R.E. Bryant (1986). Graph-based algorithms for Boolean function manipulation. *IEEE Transactions on Computers*, **C-35**(8):677–691.

[2] R.E. Bryant (1992). Symbolic Boolean manipulation with ordered binary decision diagrams. *ACM Computing Surveys*, **24**(3):293–318.

[3] R.E. Bryant (1991). On the complexity of VLSI implementations and graph representations of Boolean functions with application to integer multiplication. *IEEE Transactions on Computers*, **40**(2):205–213.

[4] M. Fujita, H. Fujisawa, and N. Kawato (1988). Evaluation and improvements of Boolean comparison method based on binary decision diagrams. In *Proceedings of the IEEE International Conference on Computer-Aided Design*, pp. 2–5. IEEE Computer Society Press.

[5] S. Malik, A. Wang, R. Brayton, and A. Sangiovanni-Vincentelli (1988). Logic verification using binary decision diagrams in a logic synthesis environment. In *Proceedings of the IEEE International Conference on Computer-Aided Design*, pp. 6–9. IEEE Computer Society Press.

[6] R. Rudell (1993). Dynamic variable ordering for ordered binary decision diagrams. In *Proceedings of the IEEE International Conference on Computer-Aided Design*, pp. 42–47. IEEE Computer Society Press.

[7] K. S. Brace, R. L. Rudell, and R. E. Bryant (1990). Efficient implementation of a BDD package. In *Proceedings of the 27th IEEE/ACM Design Automation Conference*, pp. 40–45. IEEE Computer Society Press.

[8] M. Garey and D. Johnson (1979). *Computers and Intractability: A Guide to the Theory of NP-Completeness*. W. H. Freeman.

[9] M. Davis and H. Putnam (1960). A computing procedure for quantification theory. *Journal of the ACM*, **7**(3):201–215.

[10] M. Davis, G. Logemann, and D. Loveland (1962). A machine program for theorem-proving. *Communications of the ACM*, **5**(7):394–397.

[11] J. Marques-Silva and K. Sakallah (1999). GRASP: a search algorithm for propositional satisfiability. *IEEE Transactions on Computers*, **48**(5):506–521.

[12] R. Bayardo and R. Schrag (1997). Using CSP lookback techniques to solve real-world SAT instances. In *Proceedings of the National Conference on Artificial Intelligence*, pp. 203–208.

[13] L. Zhang, C. Madigan, M. Moskewicz, and S. Malik (2001). Efficient conflict driven learning in a Boolean satisfiability solver. *Proceedings of the IEEE/ACM International Conference on Computer Aided Design*, pp. 279–285.

[14] M. Prasad, A. Biere, and A. Gupta (2005). A survey of recent advances in SAT-based formal verification. *International Journal on Software Tools for Technology Transfer (STTT)*, **7**(2):156–173.

[15] H. Zhang (1997). SATO: an efficient propositional prover. In W. McCune, ed., *Proceedings of the 14th International Conference on Automated Deduction*, Lecture Notes in Computer Science, vol. 1249, pp. 272–275. Springer.

[16] M. Moskewicz, C. Madigan, Y. Zhao, L. Zhang, and S. Malik (2001). Chaff: engineering an efficient SAT solver. In *Proceedings of the 39th ACM/IEEE Design Automation Conference*.

[17] M. H. Schulz and E. Auth (1989). Improved deterministic test pattern generation with applications to redundancy identification. *IEEE Transactions on Computer-Aided Design*, **8**(7):811–816.

[18] J. Rajski and H. Cox (1990). A method to calculate necessary assignments in algorithmic test pattern generation, *Proceedings of the International Test Conference*, pp. 25–34.

[19] W. Kunz and D. K. Pradhan (1994). Recursive learning: a new implication technique for efficient solutions to CAD problems – test, verification, and optimization. *IEEE Transactions on Computer-Aided Design*, **13**(9):1143–1158.

[20] E. Clarke, O. Grumberg, and D. Peled (1999). *Model Checking*. MIT Press.

[21] T. Cormen, C. Leiserson, R. Rivest, and C. Stein (2001). *Introduction to Algorithms*. 2nd edn. MIT Press and McGraw-Hill.

[22] E. Clarke and E. Emerson (1981). Design and synthesis of synchronization skeletons using branching time logic. In *Proceedings of Workshop on Logic of Programs*, Lecture Notes in Computer Science, vol. 131, pp. 52–71. Springer-Verlag.

[23] G. Holzmann (1991). *Design and Validation of Computer Protocols*. Prentice Hall.

[24] K. McMillan (1993). *Symbolic Model Checking: An Approach to the State Explosion Problem*. Kluwer Academic Publishers.

[25] V. Schuppan and A. Biere (2004). Efficient reduction of finite state model checking to reachability analysis. *Software Tools for Technology Transfer (STTT)*, **5**(1–2):185–204.

[26] R. E. Bryant (1992). Symbolic Boolean manipulation with ordered binary decision diagrams. *ACM Computing Surveys*, **24**(3):293–318.

[27] J. Burch, E. Clarke, D. Long, K. McMillan, and D. Dill (1994). Symbolic model checking for sequential circuit verification. *IEEE Transactions on Computer-Aided Design of Integrated Circuits and Systems*, **13**(4):401–424.

[28] T. Larrabee (1992). Test pattern generation using Boolean satisfiability. *IEEE Transactions on Computer-Aided Design of Integrated Circuits and Systems*, **11**(1):4–15.

[29] P. Abdulla, P. Bjesse, and N. Eén (2000). Symbolic reachability analysis based on SAT-solvers. In S. Graf and M. Schwartzbach, eds., *Proceedings of the 6th International Conference on Tools and Algorithms for the Construction and Analysis of Systems (TACAS)*, Lecture Notes in Computer Science, vol. 1785, pp. 411–425. Springer.

[30] P. Williams, A. Biere, E. Clarke, and A. Gupta (2000). Combining decision diagrams and SAT procedures for efficient symbolic model checking. In E. Allen Emerson and A. Prasad Sistla, eds., *Proceedings of the 12th International Conference on Computer Aided Verification (CAV)*, Lecture Notes in Computer Science, vol. 1855, pp. 124–138. Springer.

[31] A. Gupta, Z. Yang, P. Ashar, and A. Gupta (2000). SAT based state reachability analysis and model checking. In W. Hunt and S. Johnson, eds., *Proceedings of the 3rd International Conference on Formal Methods in Computer-Aided Design (FMCAD)*, Lecture Notes in Computer Science, vol. 1954, pp. 354–371. Springer.

[32] A. Biere, A. Cimatti, E. M. Clarke, M. Fujita, and Y. Zhu (1999). Symbolic model checking using SAT procedures instead of BDDs. In *Proceedings of the 36th ACM/IEEE Conference on Design Automation*, pp. 317–320.

[33] A. Biere, E. Clarke, R. Raimi, and Y. Zhu (1999). Verifying safety properties of a PowerPC microprocessor using symbolic model checking without BDDs. In N. Halbwachs and D. Peled, eds., *Proceedings of the 11th International Conference on Computer-Aided Verification (CAV)*, Lecture Notes in Computer Science, vol. 1633, pp. 60–71. Springer.

[34] F. Fallah (2002). Binary time-frame expansion. In *Proceedings of the IEEE/ACM International Conference on Computer Aided Design*, pp. 458–464.

[35] P. Bjesse, T. Leonard, and A. Mokkedem (2001). Finding bugs in an alpha microprocessor using satisfiability solvers. In G. Berry, H. Comon, and A. Finkel, eds., *Proceedings of the 13th International Conference on Computer-Aided Verification*, Lecture Notes in Computer Science, vol. 2102, pp. 454–464. Springer.

[36] F. Copti, L. Fix, R. Fraer, *et al.* (2001). Benefits of bounded model checking in an industrial setting. In G. Berry, H. Comon, and A. Finkel, eds., *Proceedings of the 13th International Conference on Computer-Aided Verification*, Lecture Notes in Computer Science, vol. 2102, pp. 436–453. Springer.

[37] N. Amla, R. Kurshan, K. McMillan, and R. Medel (2003). Experimental analysis of different techniques for bounded model checking. In H. Garavel and J. Hatcliff, eds., *Proceedings of the 9th International Conference on Tools and Algorithms for the Construction and Analysis of Systems*, Lecture Notes in Computer Science, vol. 2619, pp. 34–48. Springer.

[38] C. van Eijk and J. Jess (1995). Detection of equivalent state variables in finite state machine verification. *Proceedings of International Workshop on Logic Synthesis*, pp. 3.35–3.44.

[39] H. Cho and C. Pixley (1997). *Apparatus and Method for Deriving Correspondences between Storage Elements of a First Circuit Model and Storage Elements of a Second Circuit Model*. US patent 5 638 381.

[40] D. Anastasakis, R. Damiano, H.-K. Ma, and T. Stanion (2002). A practical and efficient method for compare-point matching. In *Proceedings of the 39th IEEE/ACM Design Automation Conference*, pp. 305–310.

[41] K. Ng, M. Prasad, R. Mukherjee, and J. Jain (2003). Solving the latch mapping problem in an industrial setting. *Proceedings of the 40th IEEE/ACM Design Automation Conference*, pp. 442–447.

[42] D. Brand (1993). Verification of large synthesized designs. In *Proceedings of the IEEE/ACM International Conference on Computer-Aided Design*, pp. 534–537.

[43] C. Berman and L. Trevillyan (1989). Functional comparison of logic designs for VLSI circuits. In *Proceedings of the IEEE/ACM International Conference on Computer-Aided Design*, pp. 456–459.

[44] SystemC. www.systemc.org/.

[45] D. G. Gajski, J. Zhu, R. Doemer, A. Gerstlauer, and S. Zhao (2000). *SpecC: Specification Language and Methodology*. Kluwer Academic.

[46] A. Stump, C. Barret, and D. Dill (2002). CVC: a cooperating validity checker. In *Proceedings of the International Conference on Computer-Aided Verification*.

[47] G. Ritter (2000). *Formal Sequential Equivalence Checking of Digital Systems by Symbolic Simulation*. Ph.D. thesis, Darmstadt University of Technology and Université Joseph Fourier.

[48] M. Weiser (1979). *Program Slices: Formal, Psychological, and Practical Investigations of an Automatic Program Abstraction*. Ph.D. thesis, University of Michigan.

[49] T. Matsumoto, H. Saito, and M. Fujita (2005). An equivalence checking method for C descriptions based on symbolic simulation with textual differences. *IEICE Transactions on Fundamentals*, **E88-A**(12):3315–3323.

[50] T. Matsumoto, H. Saito, and M. Fujita (2006). Equivalence checking of C programs by locally performing symbolic simulation on dependence graphs. In *Proceedings of International Symposium on Quality Electronic Design*, pp. 370–375.

[51] T. Matsumoto, H. Saito, and M. Fujita (2005). Equivalence checking for transformations and optimizations in C programs on dependence graphs. In *Proceedings of International Workshop on Logic and Synthesis*, pp. 357–366.

2 Transaction-level system modeling

Daniel Gajski and Samar Abdi

Model-based verification has been the bedrock of electronic design automation. Over the past several years, system modeling has evolved to keep up with improvements in process technology fueled by Moore's law. Modeling has also evolved to keep up with the complexity of applications resulting in various levels of abstraction. The design automation industry has evolved from transistor-level modeling to gate level and eventually to register-transfer level (RTL). These models have been used for simulation-based verification, formal verification, and semi-formal verification.

With the advent of embedded systems, the software content in most modern designs is growing rapidly. The increasing software content, along with the size, complexity, and heterogeneity of modern systems, makes RTL simulation extremely slow for any reasonably sized system. This has made system verification the most serious obstacle to time to market.

The root of the problem is the signal-based communication modeling in RTL. In any large design there are hundreds of signals that change their values frequently during the execution of the RTL model. Every signal toggle causes the simulator to stop and re-evaluate the state of the system. Therefore, RTL simulation becomes painfully slow. To overcome this problem, designers are increasingly resorting to modeling such complex systems at higher levels of abstraction than RTL.

In this chapter, we present *transaction-level models* (TLMs) of embedded systems that replace the traditional signal toggling model of system communication with function calls, thereby increasing simulation speed. We discuss essential issues in TLM definition and explore different classifications as well as cases for TLMs. We will also provide an understanding of the basic building blocks of TLMs. A basic knowledge of system-level design and discrete event simulation is helpful but not required for understanding TLM concepts.

2.1 Taxonomy for TLMs

Transaction-level modeling is an emerging concept that still has not been fully standardized in the industry. Different people have different notions of how TLMs

Practical Design Verification, eds. Dhiraj K. Pradhan and Ian G. Harris. Published by Cambridge University Press. © Cambridge University Press 2009.

should appear, both syntactically and semantically. This is because the original TLM definition did not provide any specific structure or semantics. However, the argument for establishing standards in TLMs is a very strong one. This is because without standards there is no possibility of sharing models, having common synthesis and analysis tools, and so on. Ad-hoc transaction-level modeling may seem attractive for having fast simulation speed for a specific design, but that approach is not conducive to establishing TLM as a viable modeling abstraction like RTL.

In 2003, a breakthrough paper on establishing taxonomy for TLMs was published. The idea was to open up the debate on what are the useful system-level models and how to position TLMs as an abstraction above RTL. The taxonomy was based on the granularity of detail in modeling the computation and communication for systems with multiple processing elements. In this section, we will present this original taxonomy with a simple example to demonstrate the differences between the proposed TLMs as well as the positioning of the TLM with respect to RTL and specification. Then we look at a different classification of TLMs based on the design objective for which the TLM is used.

2.1.1   Granularity-based classification of TLMs

In a TLM, the details of communication amongst computation components are separated from the details of computation components themselves. Communication is modeled by channels that are simply a repository for communication services. This is very similar to a class in C++. In fact, SystemC is a popular system design language that uses C++ classes to implement channels. The channel communication services are used by transaction requests that take place by calling interface functions of these channels. Unnecessary details of communication and computation are hidden in a TLM and may be added later in the design process. Transaction-level models speed up simulation and allow the exploration and validation of design alternatives at a higher level of abstraction. However, the definition of TLMs is not well understood. Without clear definition of TLMs, any predefined TLMs cannot be easily reused. Moreover, the usage of TLMs in the existing design domains, namely modeling, validation, refinement, exploration, and synthesis, cannot be systematically developed. Consequently, the inherent advantages of TLMs do not effectively benefit designers. To eliminate some ambiguity of TLMs, we attempt to define several TLMs explicitly, each of which may be adopted for a different design purpose.

To simplify the design process, designers generally use a number of intermediate models. The intermediate models slice the entire design process into several smaller design stages, each of which has a specific design objective. Since the models can be simulated and estimated, the result of each of these design stages can be independently validated. To relate different models, we introduce the system modeling graph shown in Fig. 2.1. The x-axis in the graph represents granularity of computation and the y-axis represents granularity of communication. On each axis, we have three degrees of time accuracy: untimed, approximate-timed, and cycle-timed. Untimed computation or communication represents the pure functionality of the design without any

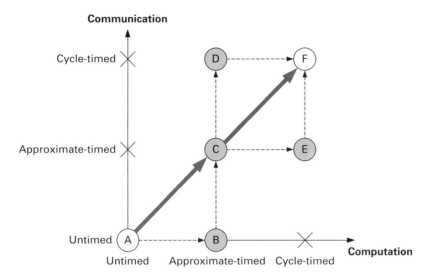

Figure 2.1   System modeling graph

implementation details. Approximate-timed computation or communication contains system-level implementation details, such as the selected system architecture and the mapping relations between tasks of the system specification to the processing elements of the system architecture. The execution time for approximate-timed computation or communication is usually estimated at the system level without cycle-accurate RTL or ISS (instruction-set simulation) level evaluation. Cycle-timed computation or communication contains implementation details at both system level and the RTL or ISS level, such that cycle-accurate estimation can be obtained.

We define six abstract models in the system-modeling graph, based on the timing granularity of computation and communication. These models, labeled A to F, are indicated on the graph by circles. Model A is the *specification* model, which has no notion of timing for either computation or communication. Model B is the *component-assembly* model, which has an approximate notion of timing for the computation part but all communication is modeled to execute in zero time. Model C is the *bus-arbitration* model, where the communication delay due to bus arbitration is factored in. Therefore, it models communication timing approximately. Model D is the *bus-functional* model, which reports accurate communication delays by factoring in both arbitration and the detailed bus protocol. However, the computation is still approximately timed. Model E is the *cycle-accurate-computation* model, which reports computation delays at the clock-cycle level of accuracy. However, the bus protocols are not modeled, which makes the communication timing only an approximation. Finally, we have model F, which is dubbed the *implementation* model because this model is traditionally the starting point for standard design tools. Both communication and computation are modeled down to the cycle-accurate level and all transactions are implemented using signal toggling according to the bus protocols. Amongst these models, the component-assembly model (B), bus-arbitration model (C), bus-functional

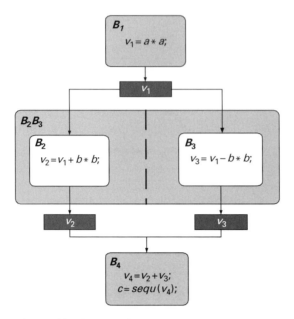

Figure 2.2   The specification model

model (D), and cycle-accurate computation model (E) are TLMs, and are indicated by shaded circles on the system modeling graph. A system-level design process takes a specification model (A) to its corresponding implementation (F). A methodology using TLMs may take any route from A to F via B, C, D, and E depending on the type of application, complexity of the platform and the focus of the verification effort. A simple methodology (A→C→F) is highlighted in Fig. 2.1. We will now delve into the modeling style and semantics of models A to F using a simple running example.

2.1.1.1   Specification model

This model captures only the system functionality and is free of any implementation details. In some literature it is also referred to as the *untimed functional model*. Figure 2.2 shows a simple specification model using graphical illustration. The round-edged rectangular boxes represent computation as a sequence of function calls or operations. We will call these computation units *behaviors*. Behaviors may also be organized into hierarchical behaviors. For example, behaviors B_2 and B_3 are composed to execute in parallel inside a hierarchical behavior B_2B_3. Behaviors that are not hierarchical are called *leaf* behaviors. These behaviors carry C code inside them that models the functionality of the behavior. Behaviors may communicate with each other using variables that are illustrated as rectangular boxes. A solid directed edge from a behavior to a variable indicates that the behavior writes to the variable. A solid directed edge from a variable to a behavior indicates that the behavior reads the variable.

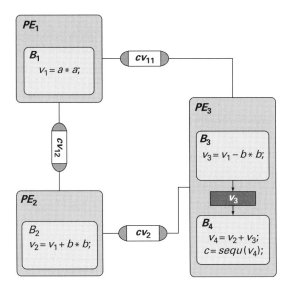

Figure 2.3   The component-assembly model

2.1.1.2   Component-assembly model

Although the specification model captures the design functionality, this is not enough for system-level verification. It is also important to verify how the design would behave when the various behaviors in the specification are distributed across different components in the platform. The verification objective is to make sure that the control and data dependencies of the behaviors are retained even if those behaviors might possibly execute in parallel on independent processing elements. This is where the component-assembly TLM comes into the picture.

In the component-assembly model, the basic modeling objects stay the same as the specification model, with the addition of the message passing channel as shown in Fig. 2.3. There is also a slight modification in the semantics of the behaviors in this model. The behaviors at the top level of the model represent concurrently executing processing elements (PEs) and global memories. This is in contrast to the specification model, where all the behaviors represent pure functionality. The mapping of behaviors in the specification model to the PEs in the platform is captured by creating an appropriate behavior hierarchy. The leaf-level behaviors of the specification model, namely B_1, B_2, B_3, and B_4 in our example, are grouped under the top-level PE behaviors. Since B_1 is mapped to PE_1, it appears under the hierarchy of PE_1, which indicates that the functionality of B_1 will be executed by PE_1. Therefore, looking at the model, we can immediately see that B_2 will be executed by PE_2 while B_3 and B_4 will be executed by PE_3.

This rearrangement of behavioral hierarchy has an impact on the communication between the behaviors as well. The PEs and memories communicate through the newly introduced message passing channels. If two behaviors with data dependence in the specification model are mapped to different PEs, then a channel must be introduced to preserve the original communication semantics. This is evident from the new channels cv_{11} and cv_{12} that replace the variables v_1 and v_2 in the original specification model. Since the writer of v_1 (B_1) is mapped to PE_1 and the reader of v_1 (B_2) is mapped to PE_2, a channel (cv_{11}) is introduced in the component-assembly model from PE_1 to PE_2, to model this data transfer between B_1 and B_2.

The system TLM at this level is a parallel composition of all the PE and memory behaviors. These TLM semantics reflect the design structure at the system level. However, the structure is modeled only for the computation part of the design, not the communication. The message-passing channels do not reflect the actual bus transactions but rather abstract the communicated data into abstract types. Although the computation structure is modeled in the component assembly model, it must be noted that the top-level behaviors are not explicitly distinguished from each other. This is because we want to keep the identity of the processing elements flexible at this time. A PE can be a custom hardware, a general-purpose processor, a DSP, or an IP. Some properties of the targeted PE may be included in the behavior model. One such key property is the approximate time it takes for the target PE to execute a certain function or operation. The estimated time of computation may be measured by profiling the code and performing system-level estimation. The estimated time is annotated into the code by inserting *wait* statements.

At this level of abstraction, any timing estimation is very coarse. For an accurate estimation, one would need to model either the micro-architecture or the finite-state machine for the PE. This would obviously slow down the simulation speed and it would take longer to evaluate the performance of the platform. This is an important trade-off that the designer must make while selecting the right TLM for his or her design space exploration. If one wants to make a coarse-grained comparative measurement of different platforms and mappings, the component-assembly model would suffice.

2.1.1.3   Bus-arbitration model

The bus-arbitration model, as the name suggests, models the communication at the bus level and also takes into account the delays resulting from arbitration over the bus. In comparison with component assembly model, channels between PEs in the bus-arbitration model represent buses, which are called abstract bus channels. Figure 2.4 shows the bus-arbitration model for our running example. The channels have three different types of interface: the master interface, the slave interface, and the arbiter interface. The channels still implement data transfer through message passing. The actual bus protocols are not modeled explicitly. Instead, the channel broadly abstracts all protocols as either blocking or non-blocking. Therefore, no cycle-accurate or pin-accurate protocol details are specified. In contrast with the point-to-point message-passing channels of the component-assembly model, the abstract bus channels have estimated approximate time delays for each transaction. This delay is incorporated into the channel methods using one *wait* statement per transaction.

The sharing of different transactions between independent PEs on the same channel poses additional modeling challenges. Since PEs are assumed to be executing

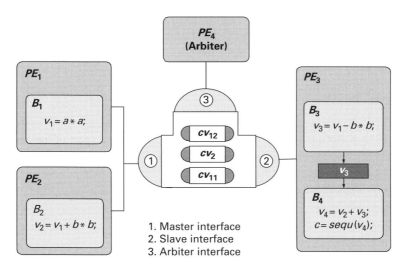

Figure 2.4 The bus-arbitration model

concurrently, independent inter-PE transactions may be attempted at the same time. The channel methods, therefore, need to distinguish between different transactions. Note that this problem would not occur in the component assembly model because the channels are point-to-point. Hence, all transactions are attempted sequentially, owing to the sequential execution semantics of leaf behaviors. The straightforward way to solve this problem in the bus-arbitration model is to use logical addresses for different transactions. As a result, the behavior calling the bus-channel method must supply the logical address of the transaction as a parameter. A simple addressing scheme would be to name each logical address the same as the original point-to-point channel. This scheme is illustrated in Fig. 2.4, where the bus channel encapsulates the original point-to-point channels in one shared entity. We can see that this mechanism of addressing is employed in all shared bus protocols, either explicitly as unique address buses or as a time-multiplexed addressing phase.

Another consequence of sharing independent transactions is the possibility of resource contention. The bus channels symbolize the physical buses in the system's bus architecture, just as the PE behaviors symbolized the computation resources in the component-assembly model. Since several different PEs connect to the same bus, there is a possibility that two transactions may be attempted in parallel. However, owing to the shared bus resources, these transactions must be serialized for correct execution. Traditionally, all shared buses implement some sort of arbitration method to perform this transaction serialization. Typically, this arbiter is a dedicated computation component that orders the bus transactions according to some specific policy. The transactions from the PEs themselves have to be modified. The PE may no longer call the communication function and expect immediate data transfer. Instead the channel function must be modified to make a request to the arbiter and wait for the grant before attempting the data transfer. Since there are two functions (*send* and *receive*) that must be executed for the transaction, one of these functions must be responsible for making the request to the arbiter. For any given transaction, the function called by the PE designated as master makes the arbitration request. The other PE must call the dual communication function as a slave. The master and slave interfaces to the bus channels are provided specifically for this reason. Methods exported by the master interface request arbitration and PEs assigned to be masters connect to this interface of the bus channel. The arbiter has its own dedicated interfaces that spool over all the arbitration requests from time to time and give the grant to the highest priority transaction based on the bus-arbitration policy. This policy is implemented inside the arbiter behavior.

2.1.1.4   Bus-functional model

The bus-functional model contains time- and cycle-accurate communication and approximate timed computation. The name comes from the fact that the detailed bus is modeled according to the specific protocol definition for both arbitration and data transfer. However, the computation part remains untouched from the bus-arbitration model. Further, we identify two types of bus-functional model, depending on the protocol definition. The *real-time-accurate* model is implemented if the protocol definition is asynchronous with timing constraints. The *cycle-accurate protocol* model is implemented when the bus protocol definition is provided on a clock cycle basis. The time-accurate model specifies the real time delay of communication, which is determined by the time diagram of the bus protocol. The cycle-accurate model can specify the time in terms of the clock cycles it takes to perform a bus read or write operation. It can easily be seen that, based on the clock cycle, a cycle accurate protocol model may be converted to a real time-accurate one. Conversely, a real time-accurate model may be converted to a cycle-accurate model in a design step called protocol refinement.

In the bus-functional model, the message-passing bus channels of the bus-arbitration model are replaced by protocol channels. Inside a protocol channel, the wires of the bus are represented by instantiating corresponding variables or signals. The communication methods inside the channel follow the detailed bus protocol by reading and writing the variables or signals that represent the bus wires. Since timing delay is associated with each such operation, the resulting model reports the timing accuracy with respect to the chosen bus protocol. At its interface, a protocol channel provides the same functions as the message passing bus channel. Therefore, there is no modification to the PE behavior code. Figure 2.5 shows the bus-functional model for our running example. As we can see, the bus signals are instantiated inside the bus channel, thereby replacing the approximate timed abstract data transfer with accurate timed protocol. The primary usage of this model is in debugging the implementation of the bus protocol.

2.1.1.5   Cycle-accurate computation model

The cycle-accurate computation model contains cycle-accurate implementations of behaviors and approximate-timed communication. This model also derives from the bus-arbitration model. In contrast with the derivation of the bus-functional model, we now leave the bus channels unchanged. However, the behaviors are modified both at

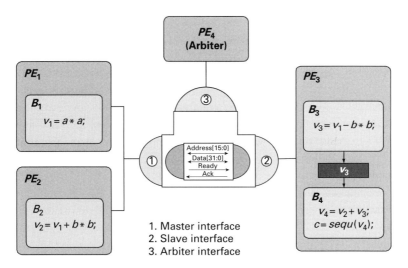

Figure 2.5   The bus-functional model

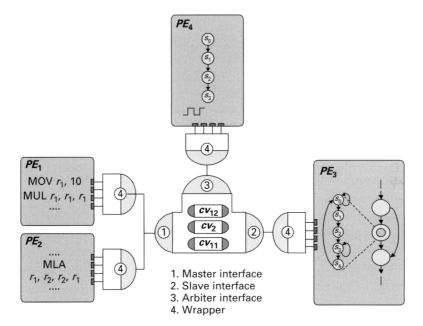

Figure 2.6   The cycle-accurate computation model

the PE level and at the leaf level. At the PE level, the behaviors are modified by replacing the abstract interface ports with pin-accurate ports. This change is visible in Fig. 2.6. The leaf-level behaviors are replaced by equivalent PE-specific SW or HW code. If the PE is a processor, then the leaf behaviors for that PE are replaced by corresponding compiled code. This replacement can be seen for PE_1 and PE_2 in Fig. 2.6, where assembly instructions have replaced the abstract C code. If the PE is a custom HW block, then a finite-state machine with data path is generated to replace the C code with synthesizable HW representation. We can see this modification in the case of PE_3 and PE_4 (the original arbiter).

Since the ports of the PE behaviors have changed, direct connection of PEs to abstract bus channels is no longer possible. To deal with this problem, a new modeling artifact, called the wrapper, is introduced into the model. These wrappers are special channels that convert data transfer from protocol-specific signal toggling on the PE's ports to abstract bus channel function calls. This mechanism functions as a bridge from the PEs to the respective bus interfaces. The primary purpose of this model is to debug the implementation of the PE. Therefore, it is possible to mix different levels of abstraction in this model. For example, if we are only interested in debugging the RTL implementation of PE_3, then PE_1, PE_2, and PE_4 may be left at the same level of abstraction as in the bus-arbitration model. In such a scenario, we would not need the wrappers for PE_1, PE_2, and PE_4. Hence, it is possible to create a high simulation speed model that is detailed only for one specific PE.

2.1.1.6   Implementation model

The implementation model has both cycle-accurate communication and cycle-accurate computation, as shown in Fig. 2.7. The components are defined in terms of their register-transfer or instruction-set architecture. Note that all the channels from the previous models have been replaced by wires at the system level. The implementation model can be derived from the bus functional model or the cycle accurate computation model. PE_1 and PE_2 are microprocessors while PE_3 and PE_4 are custom hardware units. The high-level synchronization of the bus channel is replaced by the interrupt signals and the interrupt generation and control logic. The data transfer has been replaced by protocol-specific bus interface logic and SW drivers. Essentially, all the communication functionality that was encapsulated in the bus channel methods is now incorporated in the PE behaviors. The implementation model is the model that typically serves as the input to standard EDA tools. The path from the specification model to the implementation model in the system modeling graph defines the transaction-level design methodology. We showed how the different transaction-level models can be utilized for different types of verification tasks. A sound taxonomy of TLMs and a well-defined path from specification are essential to allow overall system verification and synthesis.

2.1.2   Objective-based classification

So far we have looked at the classification of TLMs based on the modeling detail for computation and communication. As TLMs became more popular they have been employed for embedded SW development, fast performance predictions and, finally, synthesis. In the following sections we will present a classification of TLMs on the basis of modeling objectives. All contemporary TLMs are executable and are used for validation of system-level design. Also, all TLMs are constructed to allow programming and validation of embedded software. However, the focus of the modeling effort depends on the methodology used. On the one hand, designers may have a well-defined

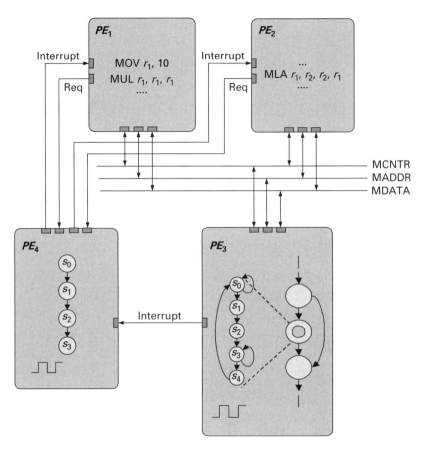

Figure 2.7 The implementation model

communication platform and only wish to tweak computation parameters for optimization. For this class of design objectives, TLMs have emerged that put the emphasis on the accuracy of estimated communication performance along with providing high simulation speed. On the other hand, platform designers may want to move to a higher level of abstraction for the sake of simplifying design specification. To fulfil this objective, it is imperative that TLMs should have well-defined synthesis semantics. The synthesis semantics and modeling rules are necessary to develop a framework where high-level abstract TLMs can be brought down to RTL and C implementations. Such low-level representation can then be easily input to traditional SW and HW tools like compilers and logic synthesizers. Another advantage of well-defined TLM semantics is the possibility of automatically generating TLMs from an abstract description of the platform, such as a graphical design input. In the following sections, we will take a look at these two modeling approaches. We will also examine whether these two approaches are entirely orthogonal or whether we have a middle ground that can give us the best of all worlds: simplified modeling, easy design input, high simulation speed, acceptable estimation, and, most importantly, a path to implementation.

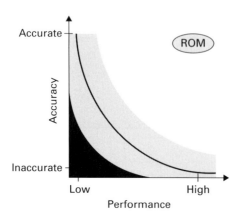

Figure 2.8   TLM trade-off

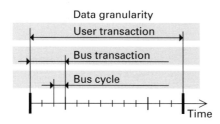

Figure 2.9   Granularity of data handling

2.2   Estimation-oriented TLMs

To explore why this trade-off (see Fig. 2.8) exists, let us examine the size of data
blocks that are typically modeled for representing abstract communication. Figure 2.9
shows how an application transfers a block of data. This block of data is called a *user
transaction* and is not restricted in size. A real bus, however, has a limitation on the
size of data blocks it can transfer. Therefore, a user transaction is broken down into
smaller elements, called *bus transactions*. A bus transaction is a bus primitive (e.g., a
store word, or store burst). Internally, the bus system needs multiple *bus cycles* to
transfer a bus transaction. For example, a bus master needs to apply the address, then
the data, and finally waits for the slave acknowledgement.

   Traditionally in TLM the designer of the bus model chooses one of the described
granularity levels. This choice dramatically influences the accuracy and performance
of the model. Modeling at the granularity of bus cycles will lead to an accurate
model. However, handling each individual bus cycle will make the model slow.
Examining the other extreme, handling complete user transaction leads to a very fast
model (since there are fewer events to be modeled), but will also be inaccurate.
Events that unexpectedly happen within a user transaction cannot be simulated by
the model.

It is intuitively understandable that a fine-grained model, which handles each individual bus cycle, will be slower than a coarse-grained model, which, for instance, handles user transactions only. However, we must ask what the main contribution to this performance penalty is. In a discrete event-simulation engine, the simulation time is advanced by wait-for-time statements. Additionally, on each call of a wait-for-time statement the simulation scheduler is executed and may context switch to another task. Therefore, there is a significant penalty involved with frequently calling a wait-for-time statement.

In the traditional TLM, the bus is modeled incrementally. The model incrementally advances the time for its granularity step (bus cycle, bus transaction, or user transaction). After each time advance it determines who can access the bus next. Therefore, a bus-cycle-based model will execute many wait-for-time statements and will be accurate but slow. On the other hand, a user-transaction-based model executes only one wait-for-time statement (one per user transaction). It will be fast, but inaccurate. It will not be able to react to any event inside the time scope of the user transaction.

2.2.1 Result-oriented modeling (ROM)

Result-oriented modeling is a modeling approach similar to TLM that hides internal states and minimizes them in order to gain execution speed. However, it does not adhere to a fixed granularity level like a traditional TLM would do. Instead of incrementally modeling a user transaction, it uses an optimistic prediction approach. Right at the beginning of the transaction, ROM calculates the total time for transferring the whole user transaction. After waiting for the predicted time, it checks whether the initial assumptions still hold true and takes corrective measurements if necessary.

2.2.2 Similarity to TLM

Like a TLM, ROM is based on the hiding of communication internals from the user. It avoids using signals and individual wires and implements data transfers by use of a single memory copy operation. In ROM, the application is only aware of the timing at the boundaries of a user transaction. All activities of the bus model within the user transaction are hidden from the communicating parties. The callers of the channel-interface functions are not aware that the transaction is split into multiple bus transactions and cycles. They are also unaware if there are competing transactions such that arbitration is involved. In other words, all communication details are encapsulated in the ROM communication channel. A very simple intuitive interface is presented to the application code developer.

However, to get the best of both worlds, we also have to do something to speed up the simulation. The main idea for speeding up the simulation is to replace the sequence of wait operations with one single wait-for-time statement. Reducing the number of wait operations is the biggest contributor to increased execution performance. This avoids running the scheduling algorithm in the simulation engine and, thus, also reduces the number of possible context switches.

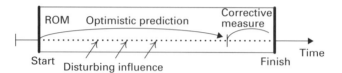

Figure 2.10   Data transfer in ROM

2.2.3   Optimistic modeling

We now look at how the transaction delay is predicted. Between the start and end times of a user transaction, the ROM can freely rearrange or omit internal events and state changes in order to eliminate costly context switches in the simulator. Instead of distributing individual wait-for-time statements to different phases of the transaction (e.g., arbitration, address, and data phase), it dynamically calculates the total time for a user transaction.

Figure 2.10 illustrates a data transfer in the ROM methodology. Note that no bus cycles are shown between the start and the end of the user transaction. Instead, the ROM implements an optimistic approach. We define the start time to be the simulated time when the application requests a user transaction. At this instant, the ROM channel makes an *optimistic prediction*. It calculates the earliest finish time for this transfer, taking into account the current state of the bus. The ROM channel then waits for the initial predicted time. During the wait time, another higher priority application process may access the bus, which may cause the transaction to take longer than initially predicted. This is because the transaction from the original lower priority process may be preempted. The ROM method records such an access as a *disturbing influence*. This is indicated by the diagonal arrows in Fig. 2.10.

After the initially calculated time has passed, the ROM verifies whether any disturbing influence has occurred (i.e., some higher priority transfer preempted the current transfer). If no disturbing influence is found, the transaction is complete. Note that in this best-case scenario, ROM uses only a single wait statement. In that sense it is similar to the abstract bus model used in a bus-arbitration TLM.

However, if a disturbing influence is found (as shown in Fig. 2.10), ROM recalculates the time for the requested user transaction, taking the updated bus state into account. Recall that the bus-state update has occurred as a result of the preemption by a higher-priority transaction. It then takes a *corrective measure* and waits for the additional time over the initial prediction. Since the original wait time is the most optimistic, any updates would only result in further waits. In other words, ROM would never need to roll back the predicted time. This fact guarantees the feasibility of a ROM-based simulation model. With the corrective measure, an overly optimistic initial prediction is corrected so that ROM can achieve 100% accuracy.

2.2.4   Measurements

Figure 2.11 illustrates the accuracy of the AMBA AHB models in a set-up where two application processes concurrently access the bus. The average error in transaction

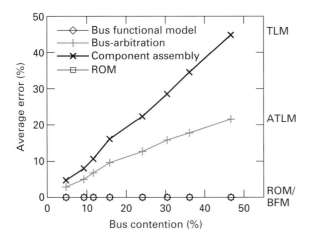

Figure 2.11   Accuracy of the AMBA AHB models

duration for the high-priority master over a varying degree of bus contention is shown. As targeted, the ROM shows 0% error for all measurements. It lies right on top of the *x*-axis. The bus-functional model, which models at a bus-cycle granularity, is accurate as well. In contrast, the traditional TLM versions (bus-arbitration TLM and component-assembly TLM) show significant error rates. These errors increase linearly with growing bus contention. At 45% contention, the component-assembly TLM reaches 45% error, making any system-timing analysis based on this TLM questionable.

To show that ROM provides high simulation speed as well, the simulation performance in a two-application process set-up was also measured. The higher-priority process produces a bus utilization of 33%. This means that the lower priority master is preempted at least one third of the time. Obviously, the more the preemption, the more corrective action needs to be taken by the ROM. Consequently, there would be a higher number of waits executed, which would result in slower simulation speed. For the set-up, the lower-priority process sends user transactions of increasing size, as shown on the *x*-axis of Fig. 2.12.

Figure 2.12 reveals the tremendous performance benefit of ROM. Both ROM and component assembly TLM are equally fast. They are three orders of magnitude faster than the bus-functional model, and one order of magnitude faster than the bus-arbitration TLM. All the models show a characteristic saw-tooth shape, owing to the non-linear split of user transactions into bus transactions. Combining the accuracy and performance measurements, we can conclude that ROM escapes the TLM trade-off for the AMBA AHB models. It is both 100% accurate and as fast as the fastest component-assembly TLM.

2.3   Synthesis-oriented TLMs

Estimation-oriented modeling is important for evaluating if a certain design's performance is satisfactory to a relatively high degree of confidence. However, it is also

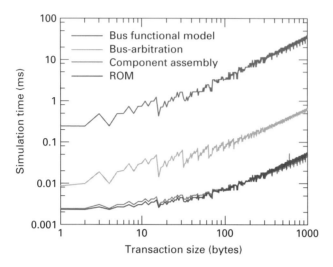

Figure 2.12   Performance of AMBA AHB models

very important to develop TLMs that can be synthesized to their respective RTL and C implementations. This modeling effort is crucial in making TLM a true next level of abstraction above the current industry standard. In this section, we give details of new research that is being done in the field of synthesizable TLMs.

The key differentiating aspect of synthesizable TLMs is that the platform objects and composition rules are clearly defined. Also, the semantics of how the transaction level platform objects map to their low-level implementations is also defined. To start with, the platform is assumed to be composed of four different types of object. These are processes, buses, memories, and transducers. Processes are similar to behaviors in that they capture the computation in the system. Buses capture the synchronization, arbitration, and data transfer between processes. Memories are storage units for addressable data and may be accessed over the bus connected to the memory. Transducers are special objects that serve as a bridge between two processes that do not share a bus. The postulation is that these objects form the necessary and sufficient set of entities needed to model a heterogeneous multiprocessor system design. From a TLM standpoint, the buses and transducers are the key objects, since they form the communication architecture of the design.

The buses are modeled as special channels, called universal bus channels (UBCs), that have well-defined templates and provide basic communication services. Processes and memories connect to UBCs in different ways. Processes use the UBC service functions to communicate amongst themselves in a rendezvous fashion. They may also use the UBC to read or write data to memories connected to the respective UBC. Memories connect to the UBC to expose their local data for access by processes connected to the UBC. Transducers are special PEs that consist of two or more specialized processes and a local first-in-first-out (FIFO) buffer. These specialized processes connect to different UBCs. Processes connected to different UBCs may send data "through" the transducer, which works in a store and forward fashion. In this section we will look at the modeling details of the UBC and the transducer and also discuss how a TLM may be constructed for a multiprocessor design with several buses, transducers, and memories. Finally, we will present TLM simulation and synthesis results for various designs of an MP3 decoder.

2.3.1   Universal bus channel (UBC)

The universal bus channel is a channel model that abstracts the system bus as a single unit of communication. The UBC provides the basic communication services of synchronization, arbitration, and data transfer that are part of a transaction. In this section, we will discuss the modeling of each of these services inside a UBC. We broadly classify transactions as either synchronized or unsynchronized. Synchronized transactions take place between two processes and require their synchronization, as the name suggests. Non-synchronized transactions (or memory transactions) are memory read-and-write operations performed by processes that require only arbitration and data transfer.

2.3.1.1   Synchronization

Synchronization is required for two processes to exchange data reliably. A sender process must wait until the receiver process is ready, and vice versa. This is essential for two reasons. Firstly, the receiver must wait until the sender has sent the data so that the received data are valid. Secondly, after sending the data, the sender must block until the data have been received so as to avoid overwriting the sent data. This type of synchronization is often referred to as double-handshake or rendezvous synchronization.

To realize rendezvous synchronization at an abstract level, we use the simple concepts of flag and event. A synchronization table is used in the UBC to keep the flags and events that are indexed by process IDs. Each unique pair of communicating processes that are connected to the same UBC has a unique <flag, event> set. These flags and events are used by a process to notify its transaction partner process that it is ready. Synchronization between two processes takes place by one process setting the flag and the other process checking and resetting the flag. Once the flag has been reset, the transacting processes are said to be synchronized.

We will refer to the process setting the flag as the *initiator* and the process resetting the flag as the *resetter*. The initiator and resetter processes for a given transaction may be determined either at compile time or at run time. If the initiator process is not fixed at compile time, then the process that becomes ready first sets the flag while the process that becomes ready second resets the flag. For example, consider that there is a transaction between processes P_1 and P_2, and assume that P_2 becomes ready first, as illustrated in Fig. 2.13. Process P_2 tests and sets the synchronization flag in a single atomic operation. When P_1 becomes ready, it attempts to test and set the flag. However, this flag is already set, so P_1 recognizes that P_2 is ready and resets the flag. Arbitration and data transfer begin once the flag is reset by the second arriving process, in this case P_1. Since the test-and-set is atomic, this scheme works even if processes become ready at the same time. Since there are several communicating processes that

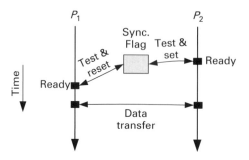

Figure 2.13   Synchronization mode decided dynamically

may share the same bus channel, a table of synchronization flags is kept in the UBC. This table is indexed by the pair of communicating processes.

If the synchronization mode for a process is to be decided dynamically, the memory storing the synchronization flag must provide atomic test-and-set operations. In many cases, this is infeasible and difficult to design. The simpler alternative is to fix the synchronization modes of the processes a priori. For example, in the earlier example, we may fix P_1 as the initiator process for this transaction and P_2 as the resetter process at compile time. This scenario is illustrated in Fig. 2.14. Hence, only P_1 can set the synchronization flag for the pair $[P_1, P_2]$. If P_2 is ready first, it must keep reading the flag until P_1 sets it.

Listing 2.1 Synchronize function in UBC

```
void Synchronize (unsigned int MyID, unsigned int PartnerID,
                  unsigned int MyMode){
   if (MyMode==UBC_INITIATOR & MyID==P_ID_P1 &&
      PartnerID==P_ID_P2){
      sync_flag_P1_P2=1;
      sync_event_P1_P2.notify();
   }
   if (MyMode==UBC_RESETTER && PartnerID==P_ID_P1 &&
      MyID==P_ID_P2){
      While (sync_flag_P1_P2 != 1){
        wait(sync_event_P1_P2);
      }
      sync_flag_P1_P2=0;
   }
   . . .
}
```

Although such a mechanism is fairly easy to implement in hardware, simulating a continuous checking of the flag by P_2 can become really slow. To avoid this problem, the continuous reading of the flag is modeled as P_2 waiting for the synchronization

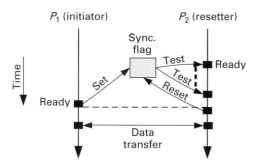

Figure 2.14   Synchronization mode decided statically

event from P_1 after the flag is set. Like the synchronization flag, the synchronization event is also defined by all pairs of processes in the synchronization table. Process P_1 notifies this event when it sets the synchronization flag. Once P_2 reads the flag as set, it recognizes that P_1 is ready and resets the flag. Arbitration and data transfer begin once the flag is reset by P_2. Listing 2.1 shows the synchronization function inside the UBC for the fixed-mode style. The body of the first *if* statement is executed when the synchronize function is called by the initiator process that sets the synchronization flag and notifies the event. The body of the second *if* statement is executed by the resetter process that checks for the status of the synchronization flag. If the flag is not yet set, it waits for the synchronization event that confirms that the flag has been set. Then it resets the flag and proceeds. The mechanism for both initiator and resetter ensures that rendezvous synchronization would always be guaranteed, irrespective of the execution speed or arrival time of the communicating processes.

   Synthesis of synchronization involves deciding the location of the synchronization flag and code generation in the communicating processes for setting and resetting the flag. For instance, in a typical interrupt-based scheme, the flag resides in the local memory of the CPU that acts as the resetter. An interrupt signal from an HW peripheral initiator sets this flag. The interface logic of the peripheral contains states that drive the interrupt signal. The interrupt controller in the CPU checks the flag every clock cycle. Alternately, the flag may reside in the local memory of the HW peripheral. Then, the flag may be set locally in the HW. The register containing the flag must have a bus address for the CPU to check the flag regularly by reading the HW register. This mechanism is called polling.

2.3.1.2   Arbitration

Since a bus is a shared resource, multiple transactions attempted at the same time must be ordered sequentially. Arbitration is modeled into the UBC to reflect such a sequential ordering of transactions. After synchronization, the resetter process attempts to reserve the bus for data transfer. This is achieved by an arbitration *request* by the resetter process. An arbitration *policy* for the bus is used to determine when this request will be *granted*. Data transfer starts following a grant. Finally, on completion of the transfer, the bus is *released* by the resetter process. The arbitration policy is now

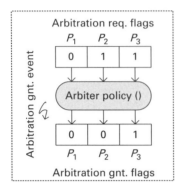

Figure 2.15   Transaction-level model of arbiter inside UBC

used to grant the bus to some other requesting process. In UBC, the arbitration features are modeled with a set of variables, events, and functions, as shown in Fig. 2.15.

All of the arbitration functionality takes place using flags and functions that are encapsulated inside the channel. None of the functions or variables is visible outside the scope of the channel. This is because the arbitration functionality is only to be used by the send and receive functions provided by the channel itself.

Arbitration requests are made by setting *arbitration request flags* indexed by the requesting process IDs. For any transaction, the resetter process sets its corresponding arbitration request flag to get permission from the arbiter to get the bus and start a transfer. In common bus-protocol terminology, the resetter is often referred to as the bus master. This is because the master controls all transactions on the bus. Arbitration grants are modeled as setting of *grant flags*, also indexed by process IDs. If a certain arbitration flag is set, it means that a bus transaction is in progress. If none of the arbitration flags is set, the bus is said to be idle or available. The setting and resetting of arbitration grant flags is done exclusively by the channel's arbiter policy function. This is done to denote which process has been granted the bus by the arbiter. Naturally, since the physical bus is a shared resource, in TLM we must ensure that at any given time only one grant flag is set per bus channel. For example the grant for P_3 alone is set although both P_2 and P_3 have requested the bus.

After making the arbiter request, the requesting process must continuously check if its grant flag is set. Only then can it proceed with the data transfer. Here we face the same problem as the continuous synchronization flag-checking problem described earlier. Even though the checking mechanism is feasible in HW, its simulation is extremely slow. We again resort to intelligent usage of events for modeling the flag-checking functionality in the TLM. We define the event *ArbitrationGranted* that is notified every time an arbiter-grant flag has been set. The event is notified after the setting of the grant flag inside the *ArbiterPolicy* function. The processes that have made an arbiter request wait for this event. Whenever the event is issued, the processes check their respective grant flags. If their grant flag is set, they proceed with data transfer, else they return to wait for the *ArbitrationGranted* event.

To set the arbiter request flag, an *ArbiterRequest* function is provided inside the UBC. This function is called by the communication service functions when the latter are called by a resetter process. The parameter to the *ArbiterRequest* function is the process ID of the requester. This function is internal to the UBC and not visible to processes connected to the UBC. Similarly, an *ArbiterGrant* function is provided that returns when the *ArbitrationGrant* flag for the respective process is set. The parameter to this function is the process ID. This function is also internal to the UBC and not visible to processes connected to the UBC. After a transaction is complete, the bus must be made available to other transactions. This responsibility of releasing the bus also rests with the resetter process. For this purpose, an *ArbiterRelease* function is provided inside the UBC for use by the communication service methods. The *ArbiterRelease* function resets the arbitration request flag for the respective process. The parameter of this function is the process ID of the calling resetter process. Just like the *ArbiterRequest* and *ArbiterGrant* functions, this function is also internal to the UBC and not visible to processes connected to the UBC.

The most important function for arbitration inside the UBC is the *ArbiterPolicy* function. This function effectively determines which process must get the bus amongst competing resetter processes. In real hardware implementation, an independent module executes the arbiter policy. Inside the UBC, however, there is no way of modeling an active process because channels are passive entities for communication whilst all the computation is done inside active processes inside PE behaviors. A simple mechanism is used to achieve the modeling of an arbiter's policy inside the passive UBC. The *ArbiterPolicy* is called by *ArbiterRequest* after setting the respective arbiter request flag. It is also called after the request flag is reset from the *ArbiterRelease* function. In effect, the arbiter evaluates the grant only when the request pattern changes. If the request pattern remains unchanged, the status quo is maintained and the arbiter is simply in an idle state. Any arbitration policy may be modeled inside the *ArbiterPolicy* function. The event *ArbitrationGranted* is notified for every execution of this function after one of the grant flags is set. The notification of this event wakes up any process that is waiting for arbitration grant. The *ArbiterPolicy* function does not take any parameters. This function is also internal to the UBC and not visible to processes connected to the UBC.

2.3.1.3 Addressing and data transfer

Data transfer is the part of a transaction where the data are copied from a sender process's memory into the memory of the receiver process. For unsynchronized transactions, transfer corresponds to reading or writing memory by a process. To distinguish between different data transfers, an addressing mechanism is used. All memories also have a range of addresses on the bus that they are connected to. Several variables and events are used to model addressing and data transfer in a UBC.

An *address table* keeps the addresses that are used by a process to transfer data between processes or between processes and memories. It is indexed by process IDs and memory ID and is local to the UBC. An example address table for a UBC connected to three processes and two memories is shown in Fig. 2.16. For

	P_1	P_2	P_3
P_1	X	0xf2	0xa0
P_2	0xf6	X	0xd2
P_3	0xfb	0xd6	X
M_1	0x10-0x80	0x10-0x60	0x60-0x80
M_2	0x8f-0xa6	0x9f-0xa2	0x8f-0xa6

Figure 2.16   Table showing the memory map of processes and memories on the UBC

unsynchronized memory transactions, the reader or writer process always sets the transaction address that is read by the memory controllers. The addressable space of a memory for a given process is given by the range in the entry with the column of the process ID and the row of the memory ID. For example, process P_2 can address memory between 0×22 and 0×48 in device M_1 (see column P_2, row M_1). To access memory on the bus, the process must be able to make an arbitration request. For synchronized transactions, the resetter process writes the transaction address. For our earlier example in Fig. 2.14, P_2 uses address $0 \times f2$ for its transaction with P_1 (column P_2, row P_1). The address values can be real bus addresses or some virtual addresses, so long as each entry for synchronized transactions is unique. Memories may have overlapping address space for multiple processes. For example, the address space of M_2 for P_2 and P_3 (row M_2, columns P_2 and P_3) is overlapping.

During the execution of the transaction, we use the *bus address* variable to store the starting address of the active transaction. This variable is set immediately after the arbitration grant and before the data transfer begins. For synchronized transactions, the resetter process writes this variable (Listing 2.2). The initiator process waits for the transaction address to be written to *BusAddress*. For memory transactions, the reader–writer process writes this variable. Memory controllers always wait for this variable to be set within the memory's address range.

Listing 2.2 Addressing by resetter in UBC

```
if (MyProcID==P_ID_P1 && ReceiverProcID==P_ID_P2)
        BusAddress=ADDR_P1_P2;
AddrSet.notify();
```

The initiator process and the memory controller processes must continuously check the value of *BusAddress* to see if they must initiate the copying of data. This leads to our usual problem, similar to the continuous flag checking needed for synchronization and arbitration. We use the usual mechanism of the event to speed up the simulation of continuous address checking. The writing of *BusAddress* is followed by the notification of event *AddressSet* by the resetter process. Listing 2.3 shows how the communication function called by process P_1 looks up the address table, sets the *BusAddress* variable to the right table entry, and notifies the *AddressSet* event.

Listing 2.3 Address checking by initiator in UBC

```
If (MyProcID==P_ID_P2 && SenderProcID==P_ID_P1){
        While (BusAddress!=ADDR_P1_P2){
                wait(AddrSet);
        }
}
```

Listing 2.3 shows the same transaction from the initiator process P_2's end. The communication function looks up the address table to find the right address that will be set by P_1 for this transaction. If the *BusAddress* variable is already set to this value, it means that the data have already been written on the bus. If not, then P_2 must wait for the *AddrSet* event before rechecking the bus address. The *AddrSet* event wakes up the initiator process to commence the data transfer.

Since the UBC is a very abstract model of the bus, we do not incorporate bus and protocol level details in the TLM. All data that are transacted in the UBC may be of arbitrary size and type. A pointer mechanism is used in the UBC to model arbitrary data-type transfer. We define a *DataPtr* variable that keeps the pointer to transacted data. This pointer is set by the sender or writer process during a transfer. The receiver or reader copies data pointed to by *DataPtr* into its local memory. Since there is no type associated with *DataPtr*, we need an extra variable to keep the size of the transacted data. The *DataSize* variable keeps the size of transacted data in bytes. It is set by the reader or writer process during a memory transaction. Memory controllers use this value to determine the number of bytes to copy. Furthermore, the memory access service in the UBC provides both read and write access to the memory connected to the bus. Therefore, along with the pointer to data and size, a flag is needed to distinguish between a read or a write transaction of the memory controller side. We use *RdWr* as a Boolean flag inside the UBC to indicate if a memory transaction is a read (0) or a write (1). This flag is written by the reader or writer process and checked by the memory during a memory transaction.

Listing 2.4 shows the *Memory Access* function provided in the UBC in a SystemC TLM. For memory transactions, the reader or writer process sets *BusAddress*. This is followed by the notification of event *AddrSet*, which wakes up the other process or memory controller that is snooping on the address bus. At the other end, the memory controller reads the address *BusAddress* to check if the address falls in its range. If the address on the bus is in range, then the memory must provide access to the right data. This is done by first computing the offset, which is the difference of the *BusAddress* and the low address of the memory. A pointer to the local memory (modeled as an array in the controller process) is needed to retrieve the correct data using the offset. If the operation is a *read* then the memory controller sets *DataPtr* to the right address in

---

Listing 2.4 MemoryAccess function in UBC called by the memory controller

```
void MemoryAccess (unsigned int MEM_LOW, unsigned int MEM_HIGH,
                   unsigned char *local_mem){
    while (1) {
        while (BusAddress < MEM_LOW | BusAddress > MEM_HIGH){
            wait (AddrSet);
        }
        if (RdWr == UBC_READ) {
            DataPtr = local_mem + (BusAddress - MEM_LOW);
        }
        else if (RdWr == UBC_WRITE) {
            memcpy (local_mem + (BusAddress - MEM_LOW), DataPtr, DataSize);
        }
        wait (BUS_DELAY, SC_NS);
    }
}
```

---

the local memory according to computed offset. If the operation is a *write*, the memory controller performs a memory copy from the data pointed to by *DataPtr* of *DataSize* number of bytes.

2.3.1.4   UBC user functions

User functions are communication service functions provided by the UBC that are visible to the user and called by processes and memories connected to the UBC. Here we summarize the user functions provided for making both synchronized and memory transactions on the bus.

1. *Send* is the method used by a process to send data to another process using synchronized transaction. The synchronization mode of the sender is selected as initiator, resetter (if determined at compile time), or either (if determined at run time). The parameters are the sender process ID, the receiver process ID, a pointer to the data being sent, the size of the data in bytes, and the synchronization mode. The receiver process must be connected to the UBC and must execute the *Recv* function for this transaction. The synchronization mode of the receiver for this transaction must be complementary to the sender. That is, if the sender mode is initiator then the receiver mode must be resetter and vice versa. If the sender mode is either, then the receiver mode must also be either.

2. *Recv* is the method used by a process to receive data from another process using synchronized transaction. The synchronization mode of the receiver is selected as initiator, resetter (if determined at compile time), or either (if determined at run time). The parameters are the receiver process ID, the sender process ID, a pointer to the location where the received data will be copied, the size of the data in bytes and the synchronization mode. The sender process must be connected to the UBC

and must execute the *Send* function for this transaction. The synchronization mode of the sender for this transaction must be complementary to the receiver. That is, if the receiver mode is initiator then the sender mode must be resetter and vice versa. If the receiver mode is either, then the sender mode must be either.

3. *Write* is the method used by a process to write data to a contiguous memory location in a non-blocking fashion. The parameters are the writer process ID, the starting memory address, a pointer to the data that need to be written, and the size of the data in bytes.

4. *Read* is the method used by a process to read data from a contiguous memory location in a non-blocking fashion. The parameters are the reader process ID, the starting memory address, a pointer to the local memory where the read data will be stored, and the size of the data in bytes.

5. *MemoryAccess* is the method used by the memory controller to service a write or read call from a process connected to the UBC. The parameters are low and high boundaries of the address range for this memory and a pointer to the start of the local memory of the device.

2.3.2   Transducer

The UBC is a sufficient communication modeling object for TLMs of processes executing on PEs that are connected to buses. However, in both homogeneous and heterogeneous systems, it is possible for several buses to exist in the system design. Furthermore, it is possible that processes that are not connected to a common bus may want to exchange data. Therefore, a special process is needed that receives data from the sender process on the first bus, stores it temporarily, and then forwards it to the receiver process over the second bus. In general, several such "bridging" processes may be needed in a dense heterogeneous system. It must be further noted that these bridging processes are used exclusively for communication and have no implication for the application itself. The only service that the application desires from the communication architecture is the reliable transmission of data from the sender process to the receiver process. To keep the modeling of communication and computation orthogonal, a generic template for the bridging logic would be immensely useful. In fact, we can define a TLM for this bridging module just as we defined the UBC for the bus. We will now introduce the transaction level modeling of the bridging module that we refer to as the transducer.

Figure 2.17 shows the simplest possible transducer. The transducer connects two buses, and its purpose is to facilitate multi-hop transactions, where one process sends data to another process that is not directly connected to the sender via a UBC. The basic functionality of the transducer is simply to receive data from the sender process, store it locally, and send it to the receiver process once the latter becomes ready. In the illustrated example, the transducer models two separate processes, IF_1 and IF_2, which interface the PEs on Bus_1 and Bus_2 to the local *FIFO*. We assume that the flow of data is from left to right. That is, PE_1 uses Bus_1 and interface IF_1 to write to the FIFO, while PE_2 uses Bus_2 and interface IF_2 to read from the FIFO. Notice the two ready signals

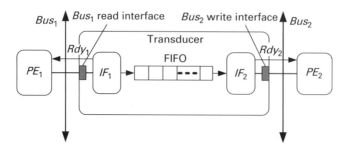

Figure 2.17   A simple transducer module

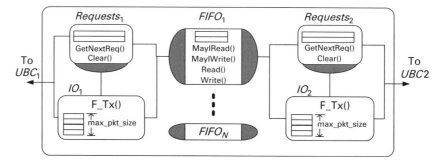

Figure 2.18   TLM template for transducer

emanating from the two interfaces. The Rdy_1 signal is triggered when the FIFO is full to indicate that PE_1 must wait for some space to be vacated in the FIFO before writing any further. Conversely, Rdy_2 signals PE_2 if the FIFO is empty, so PE_2 must wait until the FIFO has some data. The signaling mechanism and FIFO ensure that data can be sent reliably from PE_1 to PE_2 even though they do not share a common bus.

Building on the concept of a controlled FIFO, we can generalize the transducer for any platform of any degree of complexity. Figure 2.18 shows the TLM template of such a generalized transducer. The top-level behavior is connected to two different buses that are modeled by channels UBC_1 and UBC_2. There may be an arbitrary number of FIFOs, depending on the storage strategy inside the transducer. For example, we may dedicate one FIFO per pair of communicating processes. On the other extreme, there may be one unified FIFO for all possible transactions with dynamic allocation of FIFO space. Therefore, the FIFO channels may be parameterized. However, they provide four standard functions as indicated in Fig. 2.18 for checking the state of the FIFO and for reading or writing data from and to it.

To communicate with processes on UBC_1 and UBC_2, a special pair of transducer processes is defined. These are labeled IO_1 and IO_2 in Fig. 2.18. The purpose of the IO behaviors is to send and receive the communicated data over the respective buses and also to write or read it on the FIFO. For this purpose, these behaviors use the Send and Recv functions of the respective UBCs. However, this is not enough if there are multiple processes on either bus. This is because IO behaviors have no way of knowing which process to expect data from or which process to send data to. To solve this ambiguity, we introduce another pair of behaviors called *Requests*₁ and *Requests*₂. The request behaviors consist of a memory that is divided into slots where each process may write its communication request. The buffer in the request behavior is partitioned and uniquely addressed for each pair of processes communicating through the transducer. The IO behavior is interfaced to the request behavior to check for any pending requests and to call the right UBC function with the relevant parameters.

2.3.2.1 FIFO buffers

The data in transit via the transducer is stored in circular buffers, modeled as FIFO channels. The number of channels in a buffer is flexible. It may be as few as one channel and as many as the total number of communication paths through the transducer. Each path through the transducer must have one buffer assigned to it, although the buffers may be shared between different paths. The higher the degree of sharing, the more complex the management of buffers inside the FIFO channel becomes. Each buffer is modeled as a *channel* and implements an *interface* that supports four functions. These interfaces are connected to the request behavior for checking FIFO status and to the IO behavior for performing read and write on the FIFO. The interface functions are:

1. *MayIWrite*, which returns true if the requested space is available in the buffer and otherwise returns false;
2. *MayIRead*, which returns true if the requested number of bytes are present in the buffer and otherwise returns false;
3. *BufferWrite*, which copies the incoming data to the buffer and updates the tail pointer;
4. *BufferRead*, which copies data from the buffer to the output and updates the head pointer.

2.3.2.2 Request behaviors

In general, before any data are sent or received to or from the transducer, a request must be made. This request is necessary to allow the transducer interface to check if the internal FIFO buffers can accommodate the data or supply it. Such a request may be included in the packet itself, but if the packet cannot fit in the FIFO then an alternative mechanism is needed to handle this scenario. Additional logic must be implemented in the transducer IO behavior to reject the packet. Also, additional functionality is needed in the sender process to detect a packet rejection and to resend it. For simplicity, we will consider the scenario where the PE writes the request, followed by synchronization and data transfer. In the case of multiple competing processes, the requests from different processes are arbitrated by the transducer request behavior. Communication with the successful requesting process is initiated.

There are two request behaviors in the transducer, one for each bus interface. The number of words per request buffer is equal to the number of communication paths through the bridge. The request buffer is modeled as any other memory module in a PE and thus has an address range on the bus. Each word in the request buffer has a unique bus address. The requesting process writes the number of bytes it expects to read or

Listing 2.5 GetNextReady function inside the transducer request behavior

```
        . . .
    if (RequestBuffer[1]){
            *Near = P_ID_P1;
            *Remote = P_ID_P2;
            *size = RequestBuffer[1];
            *TransferType = UBC_SEND;
            *Mode = UBC_RESETTER;
            if (FIFO1->May|Write (*Remote, *Near, *size) = = TRUE)
              return TRUE;
    }
    if (RequestBuffer[2]){
            *Near = P_ID_P1
            *Remote = P_ID_P2;
            *size = RequestBuffer[2];
            *TransferType = UBC_RECV;
            *Mode = UBC_RESETTER;
            if (FIFO2->May|Read (*Near, *Remote, *size) = = TRUE)
              return TRUE;
    }
        . . .
```

write into the communication path's corresponding request buffer. The request buffer is a module that supports two functions:

1. *GetNextReady* checks the request words in the buffer in a round-robin fashion. For the chosen request, it checks whether the corresponding buffer has enough data or space to complete a transaction of the requested size, calling the buffers' functions *MayIWrite* and *MayIRead*. If the FIFO status check returns *True*, it returns the request ID and path, otherwise it checks the next pending request. Listing 2.5 shows a sample snippet from the *GetNextReady* function implemented by the request behavior.
2. *ClearRequest* removes the request from the buffer by setting the size to zero.

2.3.2.3   IO behaviors

The IO module is the interface function of the transducer that talks to other processes on the bus. It starts by calling the *GetNextReady* function in the request buffer. Then, for the selected sender or receiver process, it calls the UBC receive or send function, respectively. The IO module assumes the role of the resetter if the process is the initiator, and vice versa. The data received from the sender are written to the corresponding FIFO. The data to be sent to the receiver are first read from the corresponding FIFO before calling the transducer send function. Once the requested transaction is completed, the IO behavior removes the request by calling the *Clear* function in the request behavior.

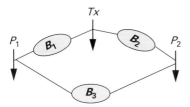

Figure 2.19   Example of design with multiple routes

	P_1	P_2
P_1	X	B_3, $B_2 - Tx - B_1$
P_2	B_3, $B_1 - Tx - B_2$	X

Figure 2.20   Global routing table

2.3.3   Routing

In a platform consisting of several buses and processes, it may be possible that two
communicating processes do not share a bus. In such cases, the transaction between the
processes must be routed through a set of buses and intermediate transducers. Such
transactions are also known as multi-hop transactions. To enable multi-hop transactions,
the TLM consists of a global routing table that is indexed by the process IDs. Each table
entry gives a set of possible routes from a source process to a destination process.

Consider the simple example of a system shown in Fig. 2.19. The design consists of two
processes P_1 and P_2. P_1 is connected to buses B_1 and B_3 while P_2 is connected to buses B_2
and B_3. A transducer is used to bridge B_1 and B_2. Therefore, to send data from P_1 to P_2,
there are two possible paths that may be chosen. Either the data may be sent directly over
B_3 or they may be sent over two hops; first from P_1 to Tx over B_1 and then from Tx to P_2
over B_2. A similar choice of reverse paths exists for sending data from P_2 to P_1. The
decision about the specific route to be selected for a given transaction may be made by the
designer. The TLM must contain the information about all possible routes between all
pairs of processes. This is important because the application developer is only concerned
about the end-to-end communication between processes. A list of routes will allow the
designer to know which particular UBC to use for a certain choice of route. This information is made available in the global routing table, as shown in Fig. 2.20.

The routing table is indexed by the pair of communicating processes. A route can be
either a bus or a string of alternating buses and processes. A route always starts and
ends with a bus. A transaction from a source process to a destination process may take
place as several transactions over intermediate hops. At each hop, the sender determines the receiver (which is either the final destination or the sender for the next hop)
by looking at the global routing table and choosing a route from itself to the

destination. If process P_1 wants to send data to P_2, it must first execute a route-selection function that returns the first bus and the first intermediate process in the route. Assume that the routing function returns B_1 and Tx. P_1 then uses the send function of B_1's UBC to send the data to Tx, by making the request for source P_1 and destination P_3. The interface of Tx connected to B_2 then reads the data from the FIFO and calls the send function of B_2 to send the data to the final destination P_2.

The routing function must be called by a process to select the appropriate UBC for the next hop. Therefore, the routing decision cannot be part of the UBC. If sending data from source to destination requires more than one transaction, the destination must send the source an acknowledgement transaction to maintain the double handshake semantics between source and destination.

2.3.4   TLMs for C-based design

So far we have looked at the building blocks of modeling the communication architecture at the transaction level. We have also looked at how to use the building blocks and processes to create TLMs. In this section we will look at the organization of TLMs in SystemC and its development from a practical standpoint.

Every processing element (PE) can consist of processes and memory elements. We can define multiple PEs in a platform, and they must be connected to at least one bus. The processing elements that contain processes have a defined internal structure, which contains C code, global functions prototypes, and SystemC code.

2.3.4.1   Processes

The processes are the C programs that execute on PEs. These programs need to interface with SystemC code in order to perform communication with other concurrently executing PEs in different PEs. Figure 2.21 shows how the code is organized in

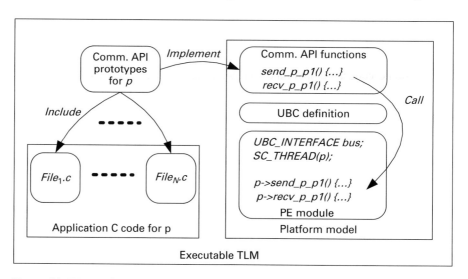

Figure 2.21   Executable TLM code organization

Listing 2.6 SystemC class encapsulating a C process

```
extern "C" int P1(void);
void *ptr_P1;
class P1: public sc_module{
  public:
      SC_HAS_PROCESS(P1);
      P1(sc_module_name name):sc_module(name){
          SC_THREAD(main);
      }
      sc_port<i_ubc> busport;
      int main(){
          ptr_p1=this;
          P1();
      }
  };
```

---

a process object. We see in Fig. 2.21 a representation of the executable TLM, which has three basic parts. First, there are the communication API prototypes, which are the global function prototypes that are included in the application C code (lower left corner). This code uses the communication APIs to access the third block: the platform model (right side of the figure). The platform model contains the communication API code that accesses SystemC code in each PE module, in order to communicate with the buses. A sample SystemC code for a process is shown in Listing 2.6.

Each process resides inside a function in the SystemC class *sc_module*. The constructor initializes all processes by defining the functions as independent *sc_thread*s. The interface to the UBC is modeled using the *sc_port* template class. The communication APIs will access this port to communicate with the bus. Inside each thread, a global pointer such as *ptr_P*$_1$ is assigned to the current object, and then the C function representing the process is called. The communication APIs exported to the application C code are global functions, which call the UBC methods inside the corresponding process's *sc_thread*. They are defined after each *sc_module*. For every process that executes with the process in this *sc_module*, a pair of communication functions is created. These are abstracted at the level of point-to-point send and receive. However, the implementation of the functions uses the relevant UBC calls.

Listing 2.7 shows the implementation of one such point-to-point send function. The function is for process P_1 to send data to process P_2. However, there is no direct bus connection between P_1 and P_2. Instead the transaction goes over a transducer. Therefore, the send transaction consists of two parts. First, a request must be written in the request buffer memory of the transducer. The address of this request is defined in the macro *ADDR_TxReq_P*$_1$*_P*$_2$ that implies that the source is P_1 and the destination is P_2. The size of the data (*r*) is written into this address in the request buffer. Then the UBC send function is called to complete the transaction from the sender's end. The

---

Listing 2.7 Implementation of point-to-point Send using UBC function calls

```
extern "C" void send_P1_P2( void *ptr, int size, int mode) {
    P1 *p = (P1*) ptr_P1;
    //Send request to transducer
    unsigned int r=size;
    p->busport->write(P_ID_P1,ADDR_TxReq_P1_P2,
                    (unsigned char*)&r,sizeof( unsigned int);
    p->busport->send(P_ID_P1,P_ID_Tx1,ptr, size, mode
                    P_ID_P1, P_ID_P2);
}
```

Listing 2.8 Implementation of point-to-point Recv using UBC function calls

```
extern "C" void recv_P_ID_P1_P2_( void *ptr, int size, int mode){
    P1 *p = (P1*)ptr_P1;
    unsigned int src= P_ID_P2;
    unsigned int dest= P_ID_P1;
    //Send request to transducer
    unsigned int r=size;
    p->busport->write(P_ID_P1,ADDR_TxReq_P2_P1,
                    (unsigned char *)&r,sizeof(unsignedint));
    p->busport->recv(P_ID_P1,P_ID_Tx1, ptr,size, mode, &src, &dest);
}
```

---

bus-port object is the port of the process P_1 that is eventually bound to the UBC at the top level instantiation of P_1.

Listing 2.8 shows, as an example, the implementation of a point-to-point receive function. The function is for process P_1 to receive data from process P_2. As mentioned earlier, there is no direct bus connection between P_1 and P_2. Instead the data must be received via a transducer. Therefore, the receive transaction consists of two parts. First, a request must be written in the request buffer memory of the transducer. The address of this request is defined in the macro $ADDR\_TxReq\_P_2\_P_1$ that implies that the source is P_2 and the destination is P_1. The size of the data (r) is written into this address in the request buffer. Then the UBC receive function is called to complete the transaction from the receiver's end.

In summary, every processing element is modeled at transaction level in SystemC as a *sc_module* with one or more *sc_thread*s, each one modeled simply using C code. For every process, there exist global point-to-point communication functions that are called by the C code of the process. These point-to-point functions are built on top of the UBC communication functions. In the case of memory elements, the *sc_module* contains an array of variables and a port to communicate with the buses. The main

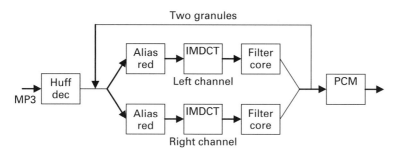

Figure 2.22   MP3 decoder algorithm

method in the *sc_module* calls the UBC *MemoryAccess* function and passes the pointer to the local array as the parameter. Other processes write and read this memory using the UBC's communication functions.

2.3.5   Synthesizable TLMs in practice: MP3 decoder design

So far we have looked at the concepts of synthesizable TLMs and practical methods for developing them in a C-based design methodology. We now take a look at some TLMs for different platforms to execute the MP3 decoder application. Figure 2.22 shows a functional block diagram of the MP3. There are five stages in the MP3 decoder. The first stage is Huffman decoding. In this stage, the MP3 bitstream is converted into 576 frequency lines, which are divided into 32 sub-bands with 18 frequency lines. For each channel in stereo mode, three functions, namely alias reduction, IMDCT, and filtercore, are executed sequentially. Each filtercore implements a fixed-point DCT function. Finally, the PCM block implements the generation of the decoded pulse-code-modulated file. Out of all the functions, the DCT in the filtercore and the IMDCT functions are the most computer intensive. Therefore, they are ideal candidates for custom implementation on HW. Also, the functions for the two different channels are data independent, so they can possibly be executed in parallel.

With the above application profile in mind, four different platforms were chosen for implementing the MP3 decoder. The first platform was simply executing everything in SW on a Microblaze processor synthesized on Xilinx board. The other three platforms we created by selectively moving the computer-intensive DCT and IMDCT functions to custom hardware blocks. This section describes the three platforms that were used to implement the MP3 decoder. It also provides experimental results that demonstrate the potential of a system design and verification methodology based around synthesizable TLMs.

Figure 2.23 shows the first example platform. On the left side of the figure, there is a Microblaze processor. The small picture in the Microblaze is the miniaturization of the MP3 decoder functional block diagram in Fig. 2.22. The black box in the Microblaze is the DCT for the left channel and it is not implemented as in software. Instead, it is implemented as a separate module in hardware and is depicted as a gray box on the

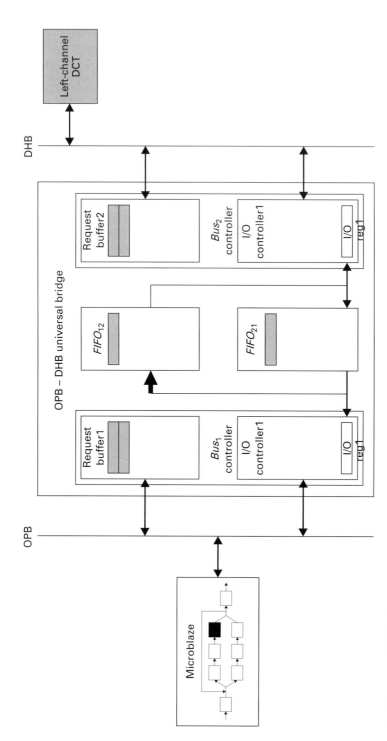

Figure 2.23 Architecture with left channel DCT in hardware

right side of the figure. The Microblaze is connected to the on-chip peripheral bus (OPB) and the hardware DCT is connected to the double-handshake bus (DHB).

To enable communication between the process mapped to the Microblaze and the DCT process mapped to HW, the OPB-DHB transducer is inserted. This transducer handles all the communication between the two buses. Inside the transducer, there are two request buffers, two I/O controllers, and two FIFOs. Two request buffers on each side of the bus have two partitions. One partition is used to store a "send request" from the OPB to the DHB and the other partition is used to store a "receive request" from the DHB to the OPB. Inside each FIFO there is only one partition. The partition in $FIFO_{12}$ is used to store the data from OPB to DHB and the other parition in $FIFO_{21}$ is used to store the data from DHB to OPB. These partitions are shown as shaded boxes in the transducer.

Figure 2.24 shows our second example. In this example, both DCTs for left and right channels are implemented in hardware. Therefore, there are two black boxes in Microblaze, and there are two DCTs on the DHB. In the transducer, there are four partitions in each request buffer. The first two request buffers are used for the left-channel DCT and the next two request buffers are used for the right-channel DCT. Each DCT uses two request buffers to store "send request" and "receive request". In each FIFO, there are two partitions. The first partition is used by the left-channel DCT and the second partition is used by the right-channel DCT. Each FIFO is unidirectional, as in the previous platform instance.

Figure 2.25 shows our third and platform example. All DCTs and IMDCTs are implemented in hardware. Therefore, there are four black boxes in Microblaze, and there are two DCTs and two IMDCTs on the DHB. As mentioned previously, one DCT or IMDCT uses two request buffers to store send and receive requests. Therefore, there are eight partitions in each request buffer. Also, one DCT or IMDCT uses one FIFO partition and there are four hardware components. As a result, there are four partitions in each FIFO.

Synthesizable TLMs using UBC for buses and transducer template for the transducers were written for each platform. We also implemented all these four platforms on Xilinx multimedia demonstration board, which uses VirtexII 2000 ff896 FPGA. The FPGA features 10 752 slices and 56 BRAMs (1008 kbits). The board-level models were written manually as well as synthesized according to the synthesis semantics of the TLM objects. Here, we present results pertaining to development time and validation time for the TLMs. We also draw a contrast in TLM-based modeling and verification versus traditional manual design.

2.3.5.1   Development time

Figure 2.26 shows the growth of time spent in modeling for different levels of abstraction. These development times are shown for the different platforms described earlier. "SW + 0" refers to a fully SW implementation. "SW + 1" refers to the second platform, where only one of the DCT blocks was implemented in HW. "SW + 2" represents the platforms with DCT for both channels implemented in HW and executed concurrently. Finally, "SW + 4" refers to the platform with both DCTs as well as IMDCTs implemented in HW.

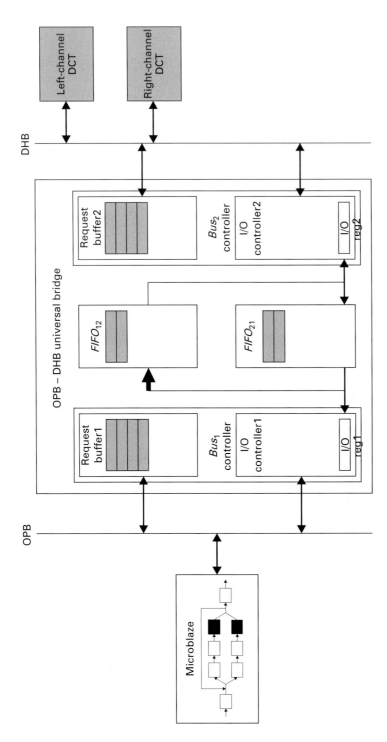

Figure 2.24 Architecture with left and right DCTs in hardware

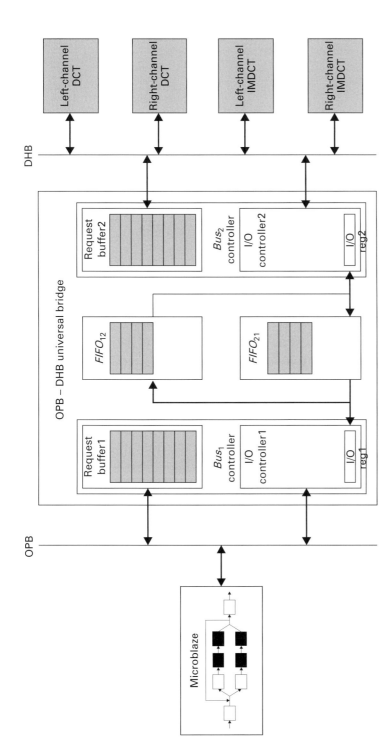

Figure 2.25 Architecture with left and right DCTs and IMDCTs in hardware

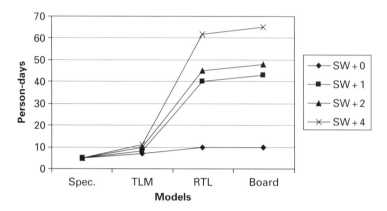

Figure 2.26   Modeling time grows sharply as we go beyond TLM

The four levels of abstraction enumerated on the *x*-axis are the specification model, the TLM in the synthesizable style, the RTL model, and the board model that was downloaded for testing on the Xilinx FPGA. A top-down design methodology was followed, where each lower-level model was derived from the next higher-level model. The development time for each abstraction level is cumulative, in the sense that the indicated development time at a given level includes the development times for models at higher levels.

A pure specification model without any implementation details took only a few person-days to code, test, and debug. It can be noted that this time is the same for all platforms. This is because the specification model does not capture any aspect of the platform, but only models the functionality. The TLMs took 6–10 days to develop, depending on the complexity of the platforms. The highest amount of time was spent in developing the RTL models because of the inherent complexity of modeling and verification in low-level hardware description languages (HDLs). Once the RTL models were finalized, it took less than three or four additional days to implement them successfully on the board. From these results, it is obvious that the bulk of the design effort is in developing the RTL models, which can be eliminated using synthesizable TLMs. This is because, using the synthesis semantics of the TLM objects, it would be possible to derive the RTL code in no time using automation tools.

2.3.5.2   Validation time

Figure 2.27 shows the validation time for simulating the four designs at different levels of abstraction. The simulation time is measured for decoding one frame of input MP3 data. The board data refers to the actual time it took for the MP3 decoder to run on the board. It can be clearly seen that the simulation speed of RTL models is a huge bottleneck in design time. The RTL simulation time is of the order of several hours compared with a few seconds for all other abstraction levels. This verification effort can be drastically minimized if there exists a reliable and proven path from the TLM to the RTL models. With synthesizable TLMs this path can be realized. As a result, most of the validation effort may be concentrated at the TLM level, where it is more feasible to develop models for different platforms and also non-prohibitive to validate them extensively.

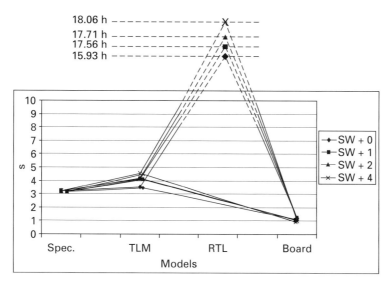

Figure 2.27   Verification time for TLMs is significantly lower than for RTL

2.4   Related work on TLMs

Transaction-level modeling has gained a lot of attention recently, ever since it was introduced [1] as part of a high-level SystemC [2,3] modeling initiative. Several use models and design flows [4,5,6] have been presented centering around TLM. In [14], the authors present semantics of different TL models based on timing granularity. Similarly, design optimization and evaluation has also been proposed using practical TLMs. [7] Generic bus architecture was defined in [8]. There have been several approaches to automatically generating executable SystemC code from abstract descriptions. [9] Modeling languages, such as UML and behavioral descriptions of systems in SystemC, have been proposed. Techniques have been proposed for the generation of SystemC TLMs from task graphs. [10] Transaction-level model generation in the SpecC [11] modeling language has been proposed for design-space exploration [12]. This method generates complex SpecC channel models for TLM in the SpecC language. Yu and Abdi have presented a novel technique and tool [13,14] for generating SystemC TLMs from application C code and graphical platforms. The proposed TLM semantics allow users to generate TLMs without understanding new system-level design languages, such as SystemC.

On the simulation front, Schirner and Dömer have proposed a result-oriented modeling (ROM) framework and methodology [15,16] that accurately models a bus transaction delay with the speed of transaction-level simulation. Siegmund and Müller [7] describe with SystemCSV an extension to SystemC and propose SoC modeling at three different levels of abstraction: physical description at the register-transfer level (RTL), a more abstract model for individual messages, and a most abstract model utilizing transactions. Their paper focuses on the interface description allowing a multilevel simulation. In [18], Caldari *et al.* describe the results of capturing the AMBA rev. 2.0 bus standard in SystemC. The bus system has been modeled at two levels of abstraction: first, a bus functional model at RTL, and second, a model at transaction-level simulating individual bus transactions. [19] The described state machine-based TLM reaches a speed-up of 100 over the RTL model. Coppola *et al.* [20] also propose abstract communication modeling. They present the IPSIM framework and show its efficient simulation. Their paper delivers a general overview of the SoC refinement and introduces their intra-module interface. Gerstlauer and Gajski describe in [21] a layered approach and propose models that implement an increasing number of International Organization for Standardization (ISO) open system interconnection (OSI) networking layers. They explain how to arrange communication and the granularity levels of simulation. Pasricha *et al.* [22] describe an approach for modeling on-chip communication architecture using TLMs. The paper introduces the concept of a model that is cycle-count accurate at transaction boundaries (CCATB) as a TLM semantics.

2.5   Summary and conclusions

In summary, we presented an overview of the state of the art in transaction-level modeling of embedded systems. We discussed the semantics of transaction-level models and provided two different types of TLM classification. The granularity-based classification positions TLMs between functional specification and pin- or cycle-accurate implementation. The concept of separating computation from communication was presented and the channel concept was introduced as a key TLM object. We also looked at objective-based classification of TLMs for estimation and synthesis. We realized that TLMs are multi-use and there is a possibility of merging these TLM definition efforts to create singular models that can be used for early design space exploration as well as inputs for a new class of system design tools that can bring these TLMs to a working implementation. Only then can TLMs become successfully adopted for IP exchange, fast verification, and globally distributed system design.

2.6   References

[1] F. Ghenassia (2005). *Transaction-Level Modeling with SystemC: TLM Concepts and Applications for Embedded Systems*. Springer.

[2] T. Grötker, S. Liao, G. Martin, and S. Swan (2002). *System Design with SystemC*. Kluwer Academic.

[3] *Open SystemC Initiative (OSCI)*. www.systemc.org.

[4] L. Cai and D. Gajski (2003). Transaction level modeling: an overview. In *Proceedings of the 1st IEEE/ACM/IFIP International Conference on Hardware/Software Codesign and System Synthesis*, pp. 19–24.

[5] A. Donlin (2004). Transaction level modeling: flows and use models. In *Proceedings of the 2nd IEEE/ACM/IFIP International Conference on Hardware/Software Codesign and System Synthesis*, pp. 75–80.

[6] K. Keutzer, A. R. Newton, J. M. Rabaly, and A. Sangiovanni-Vicentelli (2000). System level design: orthogonalization of concerns and platform-based design. *IEEE Transactions on Computer-Aided Design of Circuits and Systems*, **19**(12):1523–1543.

[7] O. Ogawa (2003). A practical approach for bus architecture optimization at transaction level. In *Proceedings of the Conference on Design, Automation and Test in Europe*, p. 20176.

[8] W. Klingauf, R. Gunzel, O. Bringmann, P. Parfuntseu, and M. Burton (2006). Greenbus – a generic interconnect fabric for transaction level modeling. In *Proceedings of the 43rd Annual Conference on Design Automation*, pp. 905–910.

[9] A. Sarmento, W. Cesario, and A. Jerraya (2004). Automatic building of executable models from abstract SoC architectures made of heterogeneous subsystems. In *Proceedings of the 15th IEEE International Workshop on Rapid System Prototyping*.

[10] D. D. Gajski, J. Zhu, R. Dömer, A. Gerstlauer, and S. Zhao (2000). *SpecC: Specification Language and Design Methodology*. Kluwer.

[11] S. Klaus, S. Huss, and T. Trautmann (2002). Automatic generation of scheduled SystemC models of embedded systems from extended task graphs. In *Proceedings of the International Forum on Design Languages*.

[12] D. Shin, A. Gerstlauer, J. Peng, R. Doemer, and D. Gajski (2006). Automatic generation of transaction-level models for rapid design space exploration. In *Proceedings of the International Conference on Hardware/Software Codesign and System Synthesis*.

[13] L. Yu and S. Abdi (2007). Automatic TLM generation for C based MPSoC design. In *Proceedings of High Level Design Validation and Test* (HLDVT).

[14] L. Yu and S. Abdi (2007). Automatic SystemC TLM generation for custom communication platforms. In *Proceedings of International Conference on Computer Design (ICCD)*.

[15] G. Schirner and R. Dömer (2005). Abstract communication modeling: a case study using the CAN automotive bus. In A. Rettberg, M. Zanella, and F. Rammig, eds., *From Specification to Embedded Systems Application*, Springer-Verlag.

[16] G. Schirner and R. Dömer (2006). Fast and accurate transaction level models using result oriented modeling. In *Proceedings of ICCAD*, pp. 363–368.

[17] R. Siegmund and D. Müller (2001). SystemCSV: an extension of SystemC for mixed multi-level communication modeling and interface-based system design. In *Proceedings of Design Automation and Test in Europe*, pp. 26–32.

[18] M. Caldari, M. Conti, M. Coppola, *et al.* (2003). Transaction-level models for AMBA bus architecture using SystemC 2.0. In *Proceedings of Design Automation and Test in Europe*, pp. 26–31.

[19] G. Schirner and R. Doemer (2006). Quantitative analysis of transaction level models for the AMBA bus. In *Proceedings of the Design Automation and Test Conference in Europe*.

[20] M. Coppola, S. Curaba, M. Grammatikakis, and G. Maruccia (2003). IPSIM: SystemC 3.0 enhancements for communication refinement. In *Proceedings of Design Automation and Test in Europe*, pp. 106–111.

[21] A. Gerstlauer and D. D. Gajski (2002). System-level abstraction semantics. In *Proceedings of International Symposium on System Synthesis*, pp. 231–236.

[22] S. Pasricha, N. Dutt, and M. Ben-Romdhane (2004). Fast exploration of bus-based on-chip communication architectures. In *Proceedings of International Conference on Hardware/Software Codesign and System Synthesis (CODES+ISSS)*, pp. 242–247.

3 Response checkers, monitors, and assertions

Harry Foster

3.1 Introduction

Functional verification is the process of confirming that an implementation has preserved the intent of the design. The intent of the design might be initially captured in an architectural or micro-architectural specification using a natural language, while the implementation might be captured as an RTL model using a hardware description language. During the verification planning process, there are three fundamental issues that must be addressed: *what* functionality of the design must be checked (observability), *how* the design is to be checked (input scenarios and stimulus), and *when* the verification process is complete (which is often defined in terms of a functional or structural coverage model). Although input stimulus generation, coverage measurement, and output checking are tightly coupled conceptually, contemporary simulation testbench infrastructures generally separate these functions into loosely coupled verification components. This chapter discusses response checking, monitors, and assertions as techniques of specifying design intent in a form amenable to verification.

3.1.1 Identifying what to check

Prior to creating response checkers, monitors, or assertions, it is necessary to identify what must be checked, which is generally part of a project's verification planning process. Figure 3.1 illustrates an abstract view of a typical design flow. The flow begins with developing a natural-language requirements document, which we refer to as an architectural specification. Next, we create an architectural model to validate the algorithmic concepts. Once validated, the architectural specification is refined; this shifts the focus from an algorithmic view of the design to a performance and feature view required for implementation. We refer to this as the micro-architectural specification, which partitions the architecture into a number of functional blocks.

Once the micro-architectural specification is approved, serious work begins on the verification plan, which traditionally has been a simulation-centric verification plan, although a verification plan might include formal verification, acceleration, emulation, and other more contemporary verification processes.

Practical Design Verification, eds. Dhiraj K. Pradhan and Ian G. Harris. Published by Cambridge University Press. © Cambridge University Press 2009.

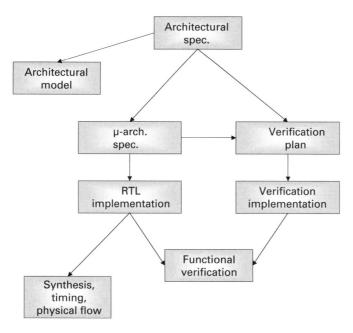

Figure 3.1   Abstract view of a typical design flow

The verification plan is the specification for the functional verification process. It typically contains the following elements:

- Overview description of the verification levels (for example, block, cluster, system),
- Architectural, micro-architectural, and implementation feature list that must be verified,
- Verification strategy (for example, directed test, constraint-random simulation, formal verification, emulation, and so forth),
- Completion criteria;
  - o Coverage goals,
  - o Bug rate curve,
  - o Verification mean time between failure.
- Resource requirements,
- Schedule details,
- Risks and dependencies.

In a contemporary verification plan, deciding what design features will be verified using static functional formal verification versus dynamic simulation-based approaches requires an understanding of design behaviors suitable for formal verfication. The following section classifies various design behaviors.

3.1.2   Classifying design behavior

Deciding whether to create a response checker, monitor, or assertion to verify a particular design feature is influenced by the target verification method we have selected during the verification planning process. For example, it might be possible to create a set of declarative assertions written in either the IEEE Std 1850–2005 Property Specification Language (PSL) [1] or IEEE Std 1800–2005 System Verilog [2] assertion language that would work in both simulation and formal verification. As an alternative, it might be significantly easier to capture a particular simulation-based data integrity check using a response checker written in C or C++. Deciding an appropriate verification method often requires an understanding of which design behaviors are best suited for a particular method. This section classifies various design behaviors and suggests appropriate verification techniques.

We can classify the behavior of today's digital systems as *reactive* or *transformational*. A reactive system is a system that continuously interacts with its environment. In other words, the current internal state of the system combined with the environment's applied input stimuli to the system determine how the system will react in terms of its next-state and output response. A classic example of a reactive system is a traffic-light controller. In fact, most controllers are reactive by definition, where their inputs arrive in endless and perhaps unexpected sequences. A transformational system, on the other hand, has all inputs ready when invoked – and an output response is produced after a certain computation period. A classic example of a transformational system is a floating-point multiplier.

We can refine our system-behavior classification into design behavior that is either *sequential* or *concurrent*.

Sequential designs, as shown in Fig. 3.2, typically operate a single stream of input data, even though there may be multiple packets at various stages of the design pipeline at any instant. An example of such sequential behavior is an instruction decode unit that decodes a processor instruction over many stages. Another example is an MPEG encoder block that encodes a stream of data. Often, the behavior of a sequential hardware block can be coded in a software language such as C or SystemC. In the absence of any other concurrent events that can interfere with the sequential computation, these blocks can be adequately verified in simulation, often validating against a C reference model.

Formal verification, on the other hand, usually faces state-explosion for sequential designs because most interesting end-to-end properties typically involve most state-holding elements of this class of design.

Concurrent designs, as shown in Fig. 3.3, deal with multiple streams of input data that collide with each other. An example is a tag generator block that serves multiple requesting agents and concurrently handles tag returns from other agents. Another example is an arbiter, especially when it deals with complex priority schemes. Both

Figure 3.2   Sequential paths

Figure 3.3 Concurrent paths

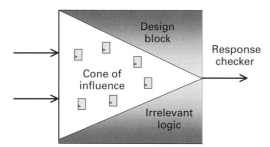

Figure 3.4 Cone-of-influence

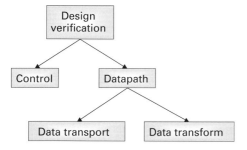

Figure 3.5 Design behavior

examples predominately have control state-holding elements in the cone-of-influence for the response checker (or assertion), as shown in Fig. 3.4.

An example of a datapath-intensive concurrent design is a switch block negotiating the traffic of packets going from multiple ingress ports to multiple egress ports. While the cone-of-influence of this type of design can have many state-holding elements, especially if the datapath is wide, a clever use of decomposition can verify correctness of one datapath bit at a time using functional formal verification (for example, model checking). This process of decomposition effectively reduces a predominantly datapath problem to a predominantly control problem. The exponential number of interesting input combinations makes achieving high simulation-based coverage challenging on concurrent designs.

Another way to characterize design blocks is as either *control* or *datapath* oriented. We can further characterize datapath design blocks as either *data transport* or *data transform*, as shown in Fig. 3.5.

Data transport blocks essentially transport packets that are generally unchanged from multiple input sources to multiple output sources, for example, a PCI express data link layer block. Data transform blocks perform a mathematical computation or an algorithm over different inputs, for example, an IFFT convolution block.

What makes data transport blocks amenable to formal verification is the independence of the bits in the datapath, often making the functional formal verification independent of the width of the datapath. Unfortunately, this kind of decomposition is usually not possible in data transform blocks.

3.1.2.1 Design behavior best suited for functional formal verification

As discussed, functional formal verification is effective for control logic and data transport blocks containing high concurrency. Examples of these blocks are: arbiters of many different kinds, on-chip bus bridges, power management units, DMA controllers, host-bus interface units, schedulers (implementing multiple virtual channels for QoS), interrupt controllers, memory controllers, token generators, credit manager blocks, and digital interface blocks (PCI express).

3.1.2.2 Design behavior better suited for simulation or emulation

In contrast, design blocks that generally do not lend themselves to functional formal verification using model checking tend to be sequential in nature and potentially involve some type of data transformation. Verifying these types of designs is generally better accomplished using simulation or emulation. Examples of designs that perform mathematical functions or involve some type of data transformation include floating-point units, graphics shading units, convolution units in DSP designs, and MPEG decoders.

3.1.3   Observability and controllability

Fundamental to the discussion of response checkers, monitors, and assertions is understanding the concepts of *controllability* and *observability*. Controllability refers to the ability to influence an embedded finite state-machine, structure, or specific line of code within the design by stimulating various input ports. While in theory a simulation testbench has high controllability of the design model's input ports during verification, it can have very low controllability of a model's internal structure. Observability, in contrast, refers to the ability to observe the effects of a specific internal finite state-machine, structure, or stimulated line of code. Thus, a testbench generally has limited observability if it only observes the external ports of the design model (because the internal signals and structures are often hidden from the testbench).

To identify a design error using the testbench approach, the following conditions must hold (that is, evaluate true):

1. The testbench must generate a proper input stimulus to activate (that is, sensitize) a bug,
2. The testbench must generate a proper input stimulus to propagate all effects resulting from the bug to a response checker or monitor attached to an output port.

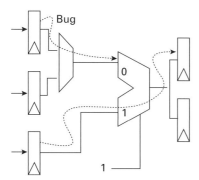

Figure 3.6   Poor observability misses bugs

It is possible, however, to set up a condition where the input stimulus activates a design error that does not propagate to an observable output port (as shown in Fig. 3.6). In these cases, the first condition cited above applies, but the second condition is absent.

Embedding internal assertions in the design model increases observability. In this way, the verification environment no longer depends on generating a proper input stimulus to propagate a bug to a response checker attached to an output port. Thus, any improper or unexpected behavior can be caught closer to the source of the bug, in terms of both time and location in the design intent.

While assertions help solve the observability challenge in simulation, they do not help with the controllability challenge. However, by adopting an assertion-based, constraint-driven simulation environment, or applying formal property checking techniques to the design assertions, we are able to address the controllability challenge.

3.2   Testbench verification components

Historically, testbenches have been monolithic programs where checkers and stimulus generators were often tightly coupled. However, today's contemporary testbenches are generally partitioned into components that are organized abstraction layers, as shown in Fig. 3.7. [3]

At the lowest level of abstraction is the design under verification (DUV), which is a signal-level, cycle-accurate model whose communication occurs through pins. Transactors (such as drivers, monitors, and responders) are abstraction converters responsible for adapting higher-level, untimed (or partially timed) transactions into a cycle-accurate sequence of interface signal values. Hence, it is unnecessary for the transaction level to understand the specific details of a DUV's interface protocols since these details are isolated within the transactors.

Figure 3.8 illustrates how various verification components might be connected in a contemporary testbench.

The details of the verification components for our contemporary testbench are given in Table 3.1.

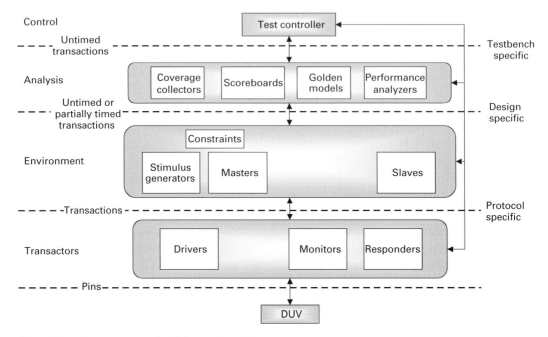

Figure 3.7   Contemporary simulation testbench layers

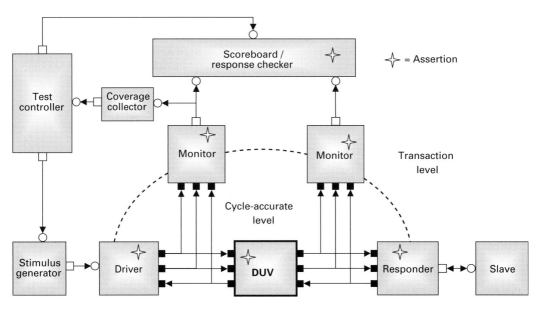

Figure 3.8   Contemporary testbench verification component interconnect

In general, testbench verification components, such as the ones illustrated in our con-temporary testbench, are written using a combination of procedural programming lan-guages and hardware description languages (for example, C/C++, SystemC, VHDL, Verilog, or SystemVerilog), as well as declarative temporal languages (for example, PSL).

Table 3.1 Contemporary testbench verification components

Testbench component	Description
DUV	Device under verification. This is a pin-level component whose function is being verified by the testbench.
Stimulus generator	This component generates transaction-level stimulus, either directed or random. It contains the algorithms and constraints used to generate directed random stimulus.
Driver	The driver converts the transaction-level stimulus into pin activations on the DUV.
Monitor	The monitor is a complement to the driver. It watches pin activity and converts it to transactions. The monitor is passive; it does not change any pins.
Responder	A responder is similar to a driver. It connects to a bus and will drive activity on the bus. The responder responds to activity on the bus rather than initiating it.
Slave	A slave is a transaction-level device whose activity is driven by the responder.
Scoreboard	A scoreboard tracks transaction-level activity from two or more devices and keeps track of information that can show if the DUV is functioning properly. For example, it might track packets in versus packets out to see if all the packets sent into a communication device made it out intact.
Response checker	A response checker is connected like a scoreboard. Rather than just tracking useful metrics, a response checker must determine whether all the responses of the DUV are correct with respect to the stimulus. In effect, it is a transaction-level model of the DUV.
Coverage collector	A coverage collector has counters organized in bins. It simply counts the transactions that are sent to it and puts the counts in the appropriate bins.
Test controller	The test controller is the decision maker within the testbench. It completes the loop from stimulus generator through driver, monitor, scoreboard, and coverage collector.
Assertion	An assertion is a specialized checker that monitors a combination of signals for correct temporal behavior. Assertions can be specified using a temporal property-specification language (for example, PSL), and are synthesized into checkers as part of the simulation environment. Alternatively, an assertion can be specified using a library of pre-synthesized assertion checkers (for example, the Accellera OVL).

3.3   Assertion-based verification

The need for an advanced verification methodology, with improved observability of design behavior and improved controllability of the verification process, has increased significantly. Over the last decade, a methodology based on the notion of "assertions" has been identified as a powerful verification paradigm that can ensure enhanced productivity, higher design quality, and, ultimately, faster time to market and higher value to engineers and end users of electronics products. [4] Assertions, as used in this context, are concise, declarative, expressive, and unambiguous specifications of desired system behavior, which are used to guide the verification process.

Assertions can be expressed using a temporal property language, such as the IEEE 1850 PSL. In addition to improving observability, using assertions results in higher design quality through:

- *Improved understanding of the design space* – resulting from the engineer's intimate analysis of the requirements, which often uncovers design deficiencies prior to RTL implementation,

- *Improved communication of design intent* among multiple stakeholders in the design process,
- *Improved verification quality* through the adoption of assertion-based verification techniques.

3.3.1   Brief introduction to SystemVerilog assertion

This section provides an overview of IEEE Std 1800–2005 SystemVerilog Assertion (SVA), which is demonstrated in Section 3.4.

A sequence is a finite series of Boolean events, where each expression represents a linear progression of time. Thus, a sequence describes a specific behavior. A SystemVerilog sequence is often described using regular expressions. This enables us to specify concisely a range of possibilities for when a Boolean expression must hold.

Sequences can be constructed as follows (where b is a Boolean expression):

b   A Boolean expression is a sequence in its simplest form,
sequence ## sequence   A sequence constructed by concatenating two sequences.

We can specify a time window with a cycle delay operation and a range.

sequence ## [range] **sequence**   A sequence constructed by concatenating two sequences.

SystemVerilog lets the user specify repetitions when defining sequences of Boolean expressions. The repetition counts can be specified as either a range of constants or a single constant expression. SystemVerilog supports three different types of repetition operator, as described in the following section.

3.3.1.1   Consecutive repetition

The consecutive repetition operator [*m:n] describes a sequence (or Boolean expression) that is consecutively repeated with one cycle delay between each repetition. For example,

$$b[^*2]$$

specifies that Boolean expression b is to be repeated for exactly two clock cycles. This is the same as specifying:

$$b \#\# 1\ b.$$

In addition to specifying a single repeat count for a repetition, SystemVerilog permits the specification of a range of possibilities for a repetition. SystemVerilog range repeat-count rules are summarized as follows:

- Each repeat count specifies a minimum and maximum number of occurrences. In the example [*m:n], m is the maximum and $n <= m$.
- The repeat count [*n] is the same as [*n:n].
- Sequences as a whole cannot be empty.

- If n is 0, then there must be either a prefix or a postfix term within the sequence specification.
- The keyword $ can be used as a maximum value within a repeat count to indicate the end of simulation. For formal verification tools, $ is interpreted as infinity (for example, [*1:$] describes a repetition of one to infinity).

3.3.1.2   Non-consecutive count repetitions

The non-consecutive count repetition operation $[*n:m]$ describes a sequence where one or more clock cycle delays are possible between the repetitions. The resulting sequence may proceed beyond the last Boolean expression occurrence in the repetition. For example,

$$a \,\#\#\, 1 \; b \, [= 1] \, \#\#\, 1 \; c$$

is equivalent to the sequence:

$$a \,\#\#\, 1 \;!\, b \, [*0 : \$] \, \#\#\, 1 \; b \,!!\,! \; b[*0 : \$] \#\#\, 1 \; c.$$

In other words, there can be any number of cycles between a and c as long as there is one b. In addition, there can be any number of cycles between a and the occurrence of b, and any number of cycles between b and the occurrence of c (that is, b is not required to precede c by exactly one cycle).

3.3.1.3   Non-consecutive exact repetitions

The non-consecutive exact repetition operator $[->n:m]$ describes a sequence where a Boolean expression is repeated with one or more cycle delays between the repetitions and the resulting sequence terminates at the last Boolean expression occurrence in the repetition. For example,

$$a \,\#\#\, b[->1] \,\#\#\, c$$

is equivalent to the sequence:

$$a \,\#\#\, 1 \;!b[*0 : \$] \#\#1 \; b \#\#1 \; c.$$

In other words, there can be any number of cycles between a and c as long as there is one b. In addition, b is required to precede c by exactly one cycle.

3.3.1.4   Sequence implication

SystemVerilog also supports operators that build complex properties out of sequences:

```
    seq₁ |-> seq₂ - sequence seq₂ starts in the last cycle of
sequence seq₁ (overlap),
    seq₁ |=> seq₂ - sequence seq₂ starts in the first cycle after
sequence seq₁
```

```
property p_mutex (clk, a, b);
  @(posedge clk)
    !(a & b);
endproperty
assert property (p_mutex(clk, enable1, enable2);
```

Figure 3.9   SystemVerilog Assertion property declaration example

3.3.1.5   Built-in functions

Assertions are commonly used to evaluate certain specific characteristics of a design implementation, such as whether a particular signal is onehot. The following system functions are included to facilitate this common assertion functionality:

- $onehot$ () returns true when one bit of a multi-bit expression is high.
- $onehot0$ () returns true when zero or one bit of a multi-bit expression is high.
- $stable$ () returns true when the previous value of the expression is the same as the current value of the expression.
- $rose$ () returns true when an expression was previously zero and the current value is non-zero. If the expression length is more than one bit, then only bit zero is used to determine a positive edge.
- $fell$ () returns true when an expression was previously one and the current value is zero. If the expression length is more than one bit, then only bit zero is used to determine a positive edge.

3.3.1.6   Declarations

SystemVerilog Assertion supports named property and sequence declarations with optional arguments, which facilitate reuse. These parameterized declarations can be referenced by name and instantiated in multiple places in designs with unique argument values. For example, we could specify the property that a and b are mutually exclusive as shown in Fig. 3.9.

3.4   Assertion-based bus monitor example

Assertions can be a very powerful tool to check the behavior of a system, but they can also be extremely useful in gathering coverage information about the transactions that have occurred. In looking at the bigger verification problem, however, a question often arises: *Is it possible to create an assertion-based bus monitor that can be used in both simulation and formal verification?* This section demonstrates how to create an assertion-based bus monitor that can be reused within a formal verification environment. The example is a simple unpipelined bus protocol, which is based on the AMBA™ Advanced Peripheral Bus (APB) protocol. [5] The goal in this section is to demonstrate the process of creating an assertion-based bus monitor, not to teach you all the details about a particular industry standard. Hence, I present a generic, simple, unpipelined-parallel-bus protocol design, which should allow you to focus on the process without getting overwhelmed by details.

Table 3.2 Signal description

Name	Summary	Description
bclk	Bus clock	The rising edge if bclk is used to time all bus transfers.
brst_n	Bus reset	An active low bus reset.
bsel	Slave select signal	These signals indicate that a slave has been selected. Each slave has its own select (for example, bsel[0] for slave 0).
ben	Strobe enable	Use to time bus accesses.
bwrite	Transfer direction	When high, write access. When low, read access
baddr[31:0]	Address	Address bus.
bdata[31:0]	Data bus	Write data driven when bwrite is high. Read data read when bwrite is low.

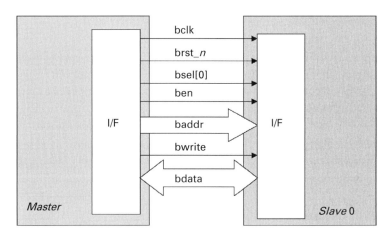

Figure 3.10   A simple, unpipelined-parallel-bus protocol design

Let us begin by examining a simple, unpipelined parallel-bus protocol design as illustrated in Fig. 3.10.

For this example, all signal transitions relate only to the rising edge of the bus clock (bclk).

Table 3.2 summarizes the bus interface signals.

We use a conceptual state machine to describe the operation of the bus for slave 0 (bsel [0]). Figure 3.11 illustrates its state diagram.

After a reset (that is, brst_n == 0), the simple parallel bus is initialized to its default INACTIVE state, which means that both bsel and ben are de-asserted. To initiate a transfer, the bus moves into the *start* state, where a slave select signal, bsel [n], is asserted by the master, selecting a single slave component (in our case, bsel [0]). The bus only remains in the *start* state for one clock cycle, and will then move to the *active* state on the next rising edge of the clock. The *active* state only lasts a

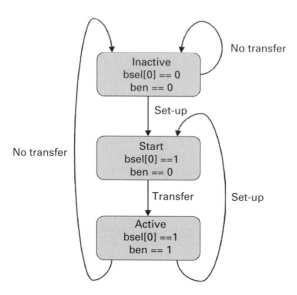

Figure 3.11   Conceptual state machine describing bus operation

single clock cycle for the data transfer. Then, the bus will move back to the *start* state if another transfer is required, which is indicated by the selection signal remaining asserted. Alternatively, if no additional transfers are required, the bus moves back to the *inactive* state when the master de-asserts the slave's select and bus enable signals.

The address (baddr[31:0]), write control (bwrite), and transfer enable (ben) signals are required to remain stable during the transition from the *start* to *active* state. However, it is not a requirement that these signals remain stable during the transition from the *active* state back to the *start* state.

3.4.1   Basic write operation

Figure 3.12 illustrates a basic write operation for the simple parallel bus interface involving a bus master and slave zero (bsel [0]).

At clock one, since both the slave select (bsel [0]) and bus enable (ben) signals are de-asserted, our bus is in an *inactive* state, as previously defined in the conceptual state machine. The state variable in the basic write operation waveform is actually a conceptual state of the bus, not a physical state implemented in the design.

The first clock of the transfer is called the *start* cycle, which the master initiates by asserting one of the slave select lines. For our example, the master asserts bsel [0], and this is detected by the rising edge of clock two. During the *start* cycle the master places a valid address on the bus and, in the next cycle, places valid data on the bus. These data will be written to the currently selected slave component.

The data transfer (referred to as the *active* cycle) actually occurs when the master asserts the bus enable signal. In our case, it is detected on the rising edge of clock three. The address, data, and control signals all remain valid throughout the *active* cycle.

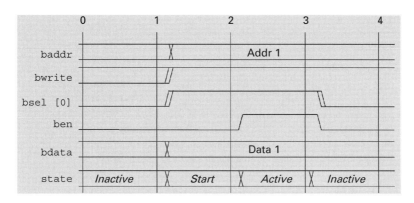

Figure 3.12   Basic write operation

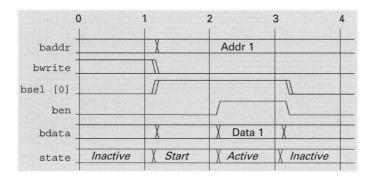

Figure 3.13   Basic read operation

When the *active* cycle completes, the bus enable signal (ben) is de-asserted by the bus master, and thus completes the current single-cycle write operation. If the master has finished transferring all data to the slave (that is, there are no more write operations), then the master de-asserts the slave select signal (for example, bsel [0]). Otherwise, the slave-select signal remains asserted, and the bus returns to the *start* cycle to initiate another write operation. It is not necessary for the address data values to remain valid during the transition from the *active* cycle back to the *start* cycle.

3.4.2   Basic read operation

Figure 3.13 illustrates a basic read operation for the simple parallel bus interface involving a bus master and slave zero (bsel [0]).

Just like the write operation, since both the slave select (bsel[0]) and bus enable (ben) signals are de-asserted at clock one, the bus is in an *inactive* state, as the conceptual state machine previously defined. The timing of the address, write, select, and enable signals are all the same for the read operation as they were for the write operation. In the case of a read operation, the slave must place the data on the bus for the master to access during the *active* cycle, which the basic read operation figure

Table 3.3 Unpipelined requirements

Property name	Summary
Bus legal state	
p_state_reset_inactive	Initial state after reset is *inactive*
p_valid_inactive_transition	*Active* state does not follow *inactive*
p_valid_start_transition	*Inactive* state does not follow *start*
p_valid_active_transition	*Active* state does not follow *active*
p_no_error_state	Bus state must be valid
Bus select	
p_bsel_mutex	Slave select signals are mutually exclusive
p_bselX_stable	Slave select signals remain stable from *start* to *active*
Bus address	
p_baddr_stable	Address remains stable from *start* to *active*
Bus write control	
p_bwrite_stable	Control remains stable from *start* to *active*
Bus data	
p_wdata_stable	Data remains stable from *start* to *active*

illustrates at clock three. Like the write operation, back-to-back read operations are permitted from a previously selected slave. However, the bus must always return to the *start* cycle after the completion of each *active* cycle.

3.4.3 Unpipelined parallel bus interface requirements

When creating a set of SVA interface assertions for the simple parallel bus, the first task is to identify a comprehensive list of natural language requirements. We begin by classifying the requirements into categories, as shown in Table 3.3.

To create a set of SystemVerilog assertions for our simple parallel bus, we begin by creating some modeling code to map the current values of the bsel and ben control signals (driven by the bus master) to the conceptual bus states. We then write a set of assertions to detect protocol violations by monitoring illegal bus state transitions, as shown in Fig 3.14.

We are now ready to write assertions for our bus interface requirements. Our first requirement states that after a reset, the bus must be initialized to an *inactive* state (which means that the bsel signals and ben are de-asserted). Hence, Fig. 3.15 demonstrates how to write a SystemVerilog property for this requirement, and then asserts the property.

Similarly, we can write assertions for all the *bus legal state* requirements as shown in Fig 3.16.

In the previous example, we created a set of properties that specifies the legal bus state transitions, as defined by our conceptual state machine.

- Property p_valid_inactive_transition specifies that if the bus is currently in an *inactive* state, then the next state of the bus must be either *inactive* again or a *start* state.

```
module unpipelined_bus_mon (
   bclk,
   brst_n,
   bsel,
   ben,
   bwrite
);
   parameter MAX_SLAVES = 4;
   input        bclk;
   input        brst_n;
   input   MAX_SLAVES-1:0] bsel;
   input        ben;
   input        bwrite;

   localparam INACTIVE   = 2'b00;
   localparam START      = 2'b01;
   localparam ACTIVE     = 2'b10;
   localparam ERROR      = 2'b11;

   wire   bus_reset    = ~brst_n;
   wire   bus_inactive = ~|(bsel) && ~ben;
   wire   bus_start    =  |(bsel) && ~ben;
   wire   bus_active   =  |(bsel) &&  ben;
   wire   bus_error    = ~|(bsel) &&  ben;
   wire [1:0] state; // conceptual state machine

   assign state        = bus_reset    ? INACTIVE :
                         bus_inactive ? INACTIVE :
                         bus_start    ? START :
                         bus_active   ? ACTIVE :
                                        ERROR ;
   // SystemVerilog Assertions here

endmodule
```

Figure 3.14   Modeling code mapping bus to conceptual states

```
property p_state_reset_inactive;
   @(posedge bclk)
      $rose(brst_n) |-> (state==INACTIVE);
endproperty
assert property (p_state_reset_inactive);
```

Figure 3.15   Bus-must-reset-to-*inactive*-state requirement

- Property p_valid_start_transition specifies that if the bus is currently in a *start* state, then on the next clock cycle, the bus must be in an *active* state.
- Property p_no_active_to_active specifies that if the bus is in an *active* state, then on the next clock cycle, the bus must be in either an *inactive* state or a *start* state.
- Finally, property p_no_error_state specifies that only valid combinations of psel and en are permitted on the bus.

For the *bus select* requirements, we can write a set of SVA properties and assert these properties as shown in Fig. 3.17.

The remaining requirements in Figure 3.18 specify that the bus controls, address, and data signals must remain stable between a bus start state and a bus active state.

```
property p_valid_inactive_transition;
  @(posedge bclk) disable iff (!rst_n)
    (state==INACTIVE) |=>
          (state==INACTIVE  ||  state==START);
endproperty
assert property (p_valid_inactive_transition);

property p_valid_start_transition;
  @(posedge bclk) disable iff (!rst_n)
    (state==START) |=> (state==ACTIVE);
endproperty
assert property (p_valid_start_transition);

property p_valid_active_transition;
  @(posedge bclk) disable iff (!rst_n)
    (state==ACTIVE) |=>
          (state==START || start==INACTIVE);
endproperty
assert property (p_valid_active_transition);

property p_no_error_state;
  @(posedge bclk) disable iff (!brst_n)
    (state!=ERROR);
endproperty
assert property (p_no error_state);
```

Figure 3.16   SystemVerilog assertions for bus legal-state requirements

```
property p_bsel_mutex;
  @(posedge bclk) disable iff (!brst_n)
    $onehot0(psel);
endproperty
assert property (p_bsel_mutex);
```

Figure 3.17   Bus-select-mutually-exclusive requirements

Although we did not explicitly state a low power requirement, many parallel buses require that the data and address lines hold their previous values to prevent switching (and, thus, consuming power). We can easily extend the properties in our previous example to check for stability between the *inactive*-to-*inactive* state, the *active*-to-*inactive* state, and the *active*-to-*start* state.

3.4.4   Unpipelined parallel bus interface coverage

The objective of adding coverage into our assertion-based monitor is to identify the proper occurrence for various types of bus transaction, which will ultimately help us analyze the quality of our simulation's random-generated stimulus. We use our assertion-based monitor to convert the bus pin-level activity to transaction streams, which are then reported to a coverage collector (see Figure 3.8). The coverage

Table 3.4 Unpipelined coverage items

Coverage item name	Summary
c_transaction	Cover types of transactions (bwrite=1 for write, bwrite=0 for read) and cover burst length (single or back-to-back transactions)

```
property p_bselX_stable;
  @(posedge bclk) disable iff (!brst_n)
    (state==START) |=> $stable(bsel);
endproperty
assert property (p_bselX_stable);

property p_baddr_stable;
  @(posedge bclk) disable iff (!brst_n)
    (state==START) |=> $stable(baddr);
endproperty
assert property (p_baddr_stable);

property p_bwrite_stable =
  @(posedge bclk) disable iff (!brst_n)
    (state==START) |=> $stable(bwrite);
endproperty
assert property (p_bwrite_stable);

property p_wdata_stable
  @(posedge bclk) disable iff (!brst_n)
    (state==START) |=> $stable(wdata);
endproperty
assert property (p_wdata_stable);
```

Figure 3.18 Bus-must-remain-stable requirements

collector can then count activity in terms of transactions or fields contained in transaction objects.

When creating a set of SVA interface assertions for our simple parallel bus, our first task is to identify a list of natural language coverage items as shown in Table 3.4.

Figure 3.19 demonstrates how to add coverage to our assertion-based monitor using a contemporary testbench analysis communication facility (for example, see AVM [3]).

The property p_transaction tracks the occurrence of (possibly consecutive) transactions on the bus. It uses the local variables *psize* and *pkind* to record the burst size and the type of the transaction. The type of transaction is recorded in the pkind *local variable* of the property when the state transitions from *inactive* to *start*, at which time the psize counter, which will record the size of the burst, is initialized to zero. As for

```
property p_transaction;
  int psize;
  tr_t pkind;

  @(posedge bclk)
    ((state==INACTIVE)
    ##1 (state==START),
            psize=0, pkind = tr_t'(bwrite))  |=>
       ((state==ACTIVE), ++psize)
    ##1 ((start==START) ##1
                  ((state==ACTIVE), ++psize) ) [*0:$]
    ##1 ((state==INACTIVE), docov(size, pkind));
  endproperty
  c_transaction: cover property (p_transaction);

  function void docov (tr_s sval, tr_t kval);
    mem_cov_t tr = new();
    tr.size = sval;
    tr.kind = kval;
    tr.trEnd = $time;
    af.write(tr); // send to analysis_fifo

  endfunction
```

Figure 3.19   SystemVerilog coverage property

the state diagram in Fig. 3.11, and the p_valid_start_transition property in Table 3.3, the state must transition from *start* to *active* on the next clock, at which time we increment the psize counter. After this, the state may, for an unspecified number of times (including zero), cycle between *start* and *active*, while incrementing *psize* in each *active* state to reflect the size of the burst. After the last *active* state, it will transition back to *inactive*. On successful completion of this property, the docov() function is called, which allows the gathered coverage information to be communicated to the rest of the testbench.

3.4.5   Analysis communication in the testbench

The simplest way to communicate coverage or other *analysis* information from a module in today's contemporary testbenches (for example, the AVM [3]) is to use the *analysis_fifo*, which is a specialized transaction-level communication object. The analysis_fifo is modeled as a class. The monitor module accesses the write() method of the fifo, which is a non-blocking function that stores the specified transaction in the fifo, from which the coverage collector will retrieve it. The monitor module is defined with an analysis_fifo port as shown in Fig. 3.20:

As you can see, SystemVerilog allows the monitor module to have an input port whose type is the parameterized analysis_fifo class. Our contemporary testbench supports connecting the "other" side of the analysis_fifo object to the coverage collector, which may itself be implemented as either a module or a class. The

```
module unpipelined_bus_mon (
   bclk,
   brst_n,
   bsel,
   ben,
   bwrite,
   analysis_fifo #(mem_cov_t) af // analysis_fifoa
);
   ...
endmodule

module top;

    analysis_fifo #(mem_cov_t) af = new("mon_fifo");

       unpipelined_bus_mon mon(.af(af) ,...);

   ...

endmodule
```

Figure 3.20   Contemporary testbench coverage analysis

`mem_cov_t` object type contains all the information needed by the coverage collector to report the transactions that occurred. In a similar manner, any of the other properties could be extended to communicate success or failure information back to the testbench as well.

3.5   Summary

In this chapter, I discussed response checking, monitors, and assertions as techniques of specifying design intent in a form amenable to verification. Functional verification is the process of confirming that the intent of the design has been preserved in its implementation. The intent of the design might initially be captured in an architectural or micro-architectural specification using a natural language, while the implementation might be captured as an RTL model using a hardware description language. During the verification planning process, there are three fundamental issues that must be addressed: *what* functionality of the design must be checked (observability), *how* the design is to be checked (input scenarios and stimulus), and *when* the verification process is complete (which is often defined in terms of a functional or structural coverage model). Although input stimulus generation, coverage measurement, and output checking are conceptually tightly coupled, contemporary simulation testbench infrastructures generally separate these functions into loosely coupled verification components. With the emergence of assertion and property language standards, such as the IEEE PSL and SVA, design teams are investigating assertion-based verification techniques. Yet there is a huge disconnection between attempting to specify an ad hoc set of assertions and implementing an effective verification flow that includes a comprehensive simulation checker. In this chapter, I demonstrated a systematic set of steps that have been found effective for creating assertion-based bus monitors that can be used within a contemporary testbench.

3.6   References

[1] *IEEE Standard for Property Specification Language (PSL)*, IEEE Std. 1850–2005.

[2] *IEEE Standard for SystemVerilog: Unified Hardware Design, Specification and Verification Language*, IEEE Std. 1800–2005.

[3] *Verification Cookbook: Advanced Verification Methodology (SystemC and SystemVerilog)*, Mentor Graphics Corporation, Version 1.2, October 28, 2005.

[4] H. Foster, A. Krolnik, and D. Lacey (2004). *Assertion-Based Design*. 2nd edn., Kluwer.

[5] ARM (1999). *AHB – AMBA Specification*, rev. 2.0. ARM.

4 System debugging strategies

Wayne H. Wolf

4.1 Introduction

Debugging embedded software is harder than debugging programs on a PC. In general-purpose software, our overriding goal is functionality or input–output behavior. Embedded systems have different and more stringent design goals than business or scientific software. Functional correctness is still a given, but it is only the first of many requirements placed on the system. These goals make embedded system debugging a very different problem.

First, embedded systems must meet real-time performance goals. Almost meeting the deadline doesn't count – a task must finish all its work by its deadline. Debugging a program for performance requires a different set of tools than is used for functional debugging. Real-time debugging is closely related to the underlying hardware architecture on which the program will execute, and so is much more closely tied to the platform than is functional debugging.

Second, many embedded systems are power and energy limited. Even embedded processors that are not powered by a battery are generally designed to power budgets to reduce heat dissipation and system cost. Like real-time performance, power and energy consumption are closely related to the hardware platform and require very different tools.

In both these cases, the characteristics and organization of the hardware platform are important determinants of the program characteristics that we want to measure and debug. When debugging programs for workstations, most of the platform dependencies that we care about come from the operating system and associated libraries. Most programmers don't worry about, for example, the details of the memory system. But debugging real-time systems must take into account all aspects of the hardware: processors, memory system, and I/O.

Several features of a platform that can influence real-time performance and power or energy consumption are:

- The pipeline introduces dependencies between nearby instructions that can cause data-dependent variations in performance and power consumption.
- The cache can have huge effects on performance. Because cache misses are expensive, cache behavior also has a profound influence on power and energy consumption.
- Bus and network contention can cause real-time performance problems.

Practical Design Verification, eds. Dhiraj K. Pradhan and Ian G. Harris. Published by Cambridge University Press. © Cambridge University Press 2009.

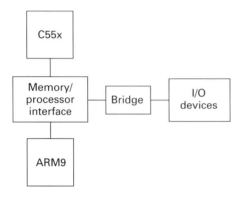

Figure 4.1   TI OMAP architecture

All this is complicated by the fact that many embedded systems are multiprocessors. Figure 4.1 shows a block diagram of the TI OMAP, which is designed for multimedia-enabled cell phones. This platform provides two processors, an ARM and a DSP, that communicate via shared memory and mailboxes. The ARM tends to perform general-purpose functions and uses the DSP as a slave to perform computationally intensive signal processing. Debugging even a two-processor system requires substantial knowledge of the middleware that governs the interactions between the processors. We should expect to see more and more embedded platforms with tens of processors and even more complex debugging processes.

4.2   Debugging tools

Debugging embedded systems requires a large set of tools. Some debugging tasks can be performed with simulator-based tools while other sorts of debugging require hardware.

Software debuggers are the base-level debugging tool. A software debugger allows the programmer to stop execution of the program, examine the state of the machine, and in some cases alter the program state before resuming execution.

Profilers can be surprisingly useful. A profiler counts the number of times that subroutines or blocks of code are executed. Profilers don't tell you details of the platform behavior, but they can be used as early warning devices to spot initial performance problems.

Simulators allow more detailed analysis of the program's behavior on a platform. Simulators provide more and easier state visibility than is possible with most platforms. A wide variety of simulators exist:

- Cache simulators simulate the state of the cache instruction-by-instruction but not all the details of the rest of the processor.
- Cycle-accurate simulators simulate the exact number of clock cycles required to execute instructions.
- Power simulators estimate the energy consumption of programs on a cycle-by-cycle basis.

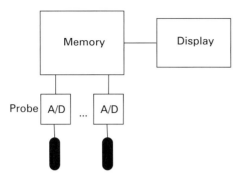

Figure 4.2 Organization of a logic analyzer

Platforms allow us to execute code with actual inputs and outputs. A platform may be an exact copy of the target system or it may be a variation. Certain types of I/O behavior are hard to simulate and much easier to test using platforms. We may not have simulators for some platforms and, therefore, be forced to debug on hardware. Platforms do generally provide less visibility than do simulators.

Logic analyzers and *oscilloscopes* can be useful for performance and power analysis. Most programmers prefer software-oriented approaches to debugging, but measuring the platform directly may uncover some transient information that would be hard to catch otherwise. Analog measurements triggered by logic analyzers can be used to make short-term power measurements.

Test generation and *test coverage* tools are often used for functional debugging. However, relatively little is known about how to generate tests or measure test coverage for real-time performance or power and energy analysis.

In-circuit emulators allow us to examine the state of a processor that is installed in a system.

4.2.1 Logic analyzers and pattern generators

Logic analyzers [1] allow a large number of digital signals to be measured simultaneously. At its most basic level, a logic analyzer is a bank of very simple oscilloscopes. The measurement probes of a logic analyzer do not record voltages as accurately as would a traditional oscilloscope, but they provide enough resolution to determine the logic level of a signal (low, high, unknown). Because the logic analyzer can record many channels at once – high-end logic analyzers can record over 100 channels – they can provide a snapshot view of the behavior of the digital system.

As shown in Fig. 4.2, a logic analyzer consists of an array of probes and conversion circuits. Trace capture is initiated by a trigger. The trigger may be simple, such as when a signal goes high. More complex triggers are Boolean combinations of signals. The most complex triggers are specified by state machines – first see one signal, then another, etc. Once the logic analyzer is triggered, data are captured periodically from a clock signal supplied by the unit under test. In some cases, a series of asynchronous events may be used to clock the trace capture. The captured traces are stored in a buffer memory for display. Logic analyzer traces must be fairly deep, at least 10 000 cycles, to be useful for most applications. Trace buffers of a million cycles deep are fairly common.

The user interface of the logic analyzer can display the traces in many different ways. Not only can the user scroll through time, but the user interface can also group and format signals. By properly specifying the user interface requirements, the user can format the trace data to look like the timing diagrams found in the processor's data sheet.

Some logic analyzers include pattern generators that can provide stimuli to a system. The pattern data are stored in a memory similar to the trace memory. The pattern generator must be triggered similarly to the triggering of the logic analyzer; in common usage the two share the same trigger. Probe circuitry applies the data to the circuit.

Modern low-cost logic analyzers are hosted by PCs. The data acquisition unit is attached to a PC that provides the user interface.

4.2.2   Power measurement

A simple way to measure power is shown in Fig. 4.3. [2] This measurement technique is designed to measure power in steady state for single instructions or very small sections of code. The CPU executes the code to be measured in a loop. An ammeter (or a voltmeter across a low-value resistor) measures the current going into the processor. The ammeter and the power supply voltage tell us the power consumption. However, care needs to be taken to determine what components are fed by the power supply tap being measured. Many modern evaluation boards provide measurement points that allow the CPU current to be measured separately from the current supplied to other components on the board. The measurement can be calibrated by also measuring the power consumption of an empty loop and comparing that value to the power consumption of the loop with the code.

More sophisticated power measurements can be made with oscilloscopes and logic analyzers. An oscilloscope can be used to measure the transient voltage across a measurement resistor. Some logic analyzers have built-in oscilloscopes that allow one or two channels to be measured with high accuracy under control of the logic analyzer. The logic analyzer's trigger can be used to determine when the oscilloscope

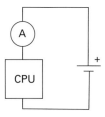

Figure 4.3   Basic power measurement

measurement is taken. This technique allows us to determine the input power over a short time interval. However, the power supply network's transient behavior must be taken into account to verify the accuracy of such measurements.

4.2.3   In-circuit emulators

An in-circuit emulator (ICE), [3] unlike a logic analyzer, allows the user to examine the internal state of a microprocessor. The in-circuit emulator replaces the stock CPU and provides facilities for program debugging, such as tracing and breakpointing. The in-circuit emulator is connected to a host PC that runs the debugging user interface. The host sends commands to the ICE and receives data from the ICE for display.

One way to build an in-circuit emulator is to build a custom module that implements the CPU architecture and provides the necessary host interface. This approach is practical only for very small, simple processors. Many modern in-circuit emulators are built using the CPU's own boundary scan system. Some CPUs put internal registers on the scan chain. Boundary scans can be used to examine register values and change them as necessary.

In-circuit emulation provides a powerful debugging interface. Most importantly, the system can be debugged on real data with real devices. Not only can the internal state of the CPU be checked; it can also be modified by scanning in a new value. This allows the user to correct problems temporarily and continue execution.

4.2.4   Emulators

Emulators and in-circuit emulators are, unfortunately, very different tools and complementary tools with very similar names. While an in-circuit emulator allows software executing on a CPU to be debugged, an emulator allows a custom hardware design to be executed before a chip is implemented.

Early emulators [4] were built from a large number of field-programmable gate arrays (FPGAs) connected in a network. A logic design could be compiled into the FPGA network. Emulators could generally execute logic designs at clock rates less than that achievable by a custom VLSI implementation, but could run much faster than an HDL simulator. The emulator could be plugged into existing hardware so that the emulated logic design could be run in its final environment. The emulator provides a number of debugging features. More recent emulators use a form of hardware-accelerated simulation to implement the logic design.

4.2.5   Profilers

A profiler gathers statistics by measuring an executing program on a standard CPU. There are two major ways to implement profiling. One method, known as PC sampling, uses a timer to interrupt the program to be profiled periodically and record the program's PC value. The other method inserts code into the program to be profiled that increments counters each time the program's execution passes certain points.

A separate program can be used as the profiled program finishes to display the statistics gathered from execution.

gprof [5] is a widely used Unix profiler. *gprof* produces trace files that are post-processed to generate reports for the user. Intel Vtune is designed to profile both single-threaded and multi-threaded applications.

4.2.6   CPU simulators

A CPU simulator is a program that implements the instruction set of a CPU. The simulator can read binary files and execute them with the same results as the CPU. The simulator is, of course, much slower than the CPU.

Simulators can be constructed at many different levels of accuracy. For embedded system debugging, the most interesting type of software simulator is the cycle-accurate simulator. This simulator not only implements the functionality of the instruction set, it also determines the number of clock cycles required to execute each instruction. Cycle-accurate simulation is not simple – instruction execution time depends on pipeline state, cache state, and data values. Cycle-accurate simulators are very complex programs; a cycle-accurate simulator may not exist for the CPU model that you are interested in.

Power simulators are cycle-accurate simulators that also estimate the power consumption of program execution. Power simulators are even more complex than cycle-accurate simulators. The power number consumption values they produce are not exact but are generally reasonable estimates.

SimpleScalar [6] is a well-known cycle-accurate simulation framework. Simple Scalar is a toolkit that allows you to create a simulator for a CPU architecture of your choice. SimpleScalar models have been constructed for a number of popular CPUs. SimplePower [7] and Wattch [8] are two well-known power simulators.

4.3   Debugging commands

Several types of debugging commands or features are useful when debugging embedded systems. Some of them are typical in general-purpose debuggers while some are unique to embedded systems.

General-purpose debuggers offer four major types of debugging command:

- Instructions can be traced. When the program counter reaches the location of a traced instruction, the debugger emits a message. Tracing allows the user to keep track of program execution.
- Breakpoints can be inserted. When the program counter reaches a breakpoint location, execution of the target program stops. Control returns to the debugger interface.
- Memory locations can be examined. These locations can be either data or instructions.
- Memory locations can be changed. Again, the changed locations can be either data or instructions.

General-purpose debuggers also provide source-level debugging. A minimal debugger would simply list the contents of a location in hex or some other base, independent of whether the location is for code or data. A more sophisticated debugger will disassemble the location to its assembly location. Source-level debugging goes one step further and relates the memory image to the original high-level language source code.

General-purpose debuggers work by modifying the target program's image in memory. Instructions in the target program are replaced by calls to the debugger. The replaced instructions must be executed separately, typically by moving the debugger call temporarily to a new location.

Specialized embedded-system debugging tools relate primarily to information about time. Traces are a common way to display execution data. While general-purpose debuggers trace only selected instructions, logic analyzers or ICEs capture traces of activity on every clock cycle. Logic analyzers and ICEs can often display traces in assembly-language format. They may allow the user to filter the trace to show only certain items, such as when a specified location is accessed. Some in-circuit emulators provide timers that can be used to measure the time required for a specified action or to control when an action takes place.

4.4   Functional debugging

The goal of functional debugging is to find and fix problems in the program's basic input and output behavior. Functional debugging does not take into account timing properties of the systems, which means that some sorts of I/O related bugs cannot be found using functional debugging techniques. However, functional debugging is an important first step – clearing out functional bugs makes it simpler to identify more subtle timing bugs.

A surprising amount of functional debugging can be done without the target platform. A great deal of code is developed on other platforms and then moved to the embedded target. This approach takes advantage of powerful debugging environments and the faster turnaround times generally given by native-only development. Code developed on another platform can and should be debugged on the initial platform before moving it to the target.

To test the code, I/O stubs must generally be developed to adapt the code to the development platform. Since the target I/O is not available, input must be provided somehow and outputs must be captured for analysis. In many cases, input traces can be captured or generated by some means and kept as files on the development system. Stub routines can then read the trace files and provide the input data to the code in the proper format. The stub routines can simulate timing constraints on inputs by aligning traces to be sure that the proper combinations of inputs are delivered together.

4.5   Performance-oriented debugging

How do we know that we missed a performance or power requirement? A good place to start is by profiling the program. Profiling only counts the number of times that source code units of function – subroutines or lines of code – are executed. However, this is often enough to help us find the big performance problems. Profiling is easy, quick, and doesn't require the platform in either simulated or real form. We can save more detailed methods for debugging problems that require them.

A fundamental choice when debugging the system is whether to debug the platform or to debug a simulation. Larger examples can generally be run on the platform, since the I/O devices are available. However, it may be hard to breakpoint or otherwise observe the required behavior on the platform. In many cases, a combination of platform and simulator debugging runs may be required to isolate and fix the problem.

A timing bug is the failure of a system to generate an output at the required time. Timing bugs can be caused by many different errors. Let us consider them one at a time.

A single process or program may take too long to execute. In this case, an in-circuit emulator or CPU simulator should help debug the program. Profiling may be able to help isolate the problem as well.

The cache behavior of a program may cause timing problems. Cache behavior is more complex to debug than simple performance problems because it often depends on program state. The contents of both the data and instruction cache depend on input data values – instruction behavior may depend upon data values. Therefore, the program may exhibit the timing bug under some circumstances but not others. Furthermore, in banked memory systems, placement of data in memory may cause timing problems.

The behavior of I/O devices may influence timing. Subtle changes in the behavior of asynchronous devices may cause timing problems that influence the behavior of a program. Consider, for example, what happens when a device driver takes too long to execute. In a priority-driven interrupt system, the driver may prevent another lower-priority driver from responding, either delaying the data from that device or causing it to be dropped totally.

Multitasking and the real-time operating system (RTOS) may cause timing problems or amplify other timing problems. For example, if a high-priority process takes too long to execute, it may delay the execution of a lower-priority process. This problem is ultimately fixed by reducing the execution time of the faulty process, but the problem must be traced through the RTOS to be fully understood.

Unfortunately, relatively little is known about causal analysis for performance. We don't have good ways to generate input vector sets that are likely to expose timing bugs. As a result, we simply have to rely on functional testing to uncover timing problems. Because the test sequences aren't designed to uncover testing problems, they don't give us much information as to the possible causes of the timing problems. Only careful debugging can be used to find the root causes and possible cures for timing bugs.

4.6   Summary

The bad news is that embedded systems provide the opportunity for a much wider repertoire of bugs than do general-purpose programs. Embedded systems can exhibit a wide variety of timing and power consumption bugs in addition to typical functional bugs.

The good news is that embedded system programmers have a wide range of tools available to help them. In-circuit emulators, simulators, and other tools can help expose system state in ways that can clarify the behavior of timing and power bugs.

4.7   References

[1] R. A. Witte (1993). *Electronic Test Instruments: Theory and Applications*, PTR Prentice Hall.

[2] V. Tiwari, S. Malik, and A. Wolfe (1994). Power analysis of embedded software: a first step toward software power minimization. *IEEE Transactions on VLSI Systems*, **2**(4):437–445.

[3] J. G. Ganssle (1999). ICE technology unplugged. *Embedded Systems Programming*, **12**(11). www.embedded.com/1999/9910/9910sr.htm.

[4] S. P. Sample, M. R. D'Amore, and T. S. Payne (1992). *Apparatus for Emulation of Electronic Hardware System*. US patent 5 109 353.

[5] J. Fenlason and R. Stallman (1998). *GNU gprof: The GNU Profiler*. www.cs.utah.edu/dept/old/texinfo/as/gprof_toc.html.

[6] SimpleScalar LLC. http://www.simplescalar.com.

[7] W. Ye, N. Vijaykrishna, M. Kandemir, and M. J. Irwin (2000). The design and use of SimplePower: a cycle-accurate energy estimation tool. In *Proceedings of the Design Automation Conference*. ACM Press.

[8] D. Brooks, V. Tiwari, and M. Martonosi (2000). Wattch: a framework for architectural-level power analysis and optimizations. In *27th International Symposium on Computer Architecture*.

5 Test generation and coverage metrics

Ernesto Sánchez, Giovanni Squillero, and Matteo Sonza Reorda

5.1 Introduction

Digital circuits are usually produced following a multi-step development process composed of several intermediate design phases. Each one is concluded by the delivery of a model that describes the digital circuit in increasing detail and with different abstraction levels. The first design step usually produces the highest abstraction level model, which describes the general behavior of the circuit leaving internal details out; whereas the last steps provide lower-level descriptions, with more detail and closer to the actual implementation of the circuit. Clearly, the lower the abstraction level, the higher the complexity of the resulting model.

In the following, some of the main characteristics of the most commonly adopted design abstraction levels as well as the main features of the delivered models at each level will be sketched. It is important to note that levels of abstraction higher or lower than those described here could also exist in a design cycle; but we only focus on the most commonly adopted ones.

- *Architectural level*
  This is often the highest abstraction level: the circuit model delivered here is used as a reference since it contains few implementation details. The main goal at the architectural level is to provide a block architecture of the circuit implementing the basic functional specifications. The delivered model is usually exploited to evaluate the basic operations of the design and the interactions among the components within the system. At this design level, a complete simulatable model may be built in some high-level language; typically, these models do not contain timing information.
- *Register transfer level (RTL)*
  Models delivered at this level contain all functional details of the design, together with accurate cycle-level timing information. At this level, clocked storage elements become visible, and (as the name recalls) these storage elements are mainly registers. However, the resulting models do not include detailed timing information, such as propagation delays of each block.

Practical Design Verification, eds. Dhiraj K. Pradhan and Ian G. Harris. Published by Cambridge University Press. © Cambridge University Press 2009.

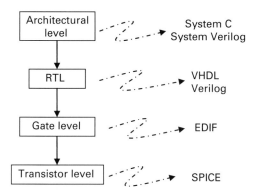

Figure 5.1   Abstraction levels of a typical design cycle

- *Gate level*

  At this level the design is described in terms of logic gates. All the interconnections between different elements within the design are thoroughly detailed, as well as all individual logic gates. Complex designs at this level can be difficult to simulate, owing to the high amount of available information that the models contain (e.g., a 16×16 multiplier, described in no more than 20 lines using a hardware description language, could count on about 2.5k gates). This level is still significantly abstract because there is no information about actual transistors.

- *Transistor level*

  This level is often not considered as an *abstraction* level of the circuit because the delivered model entirely represents the design in terms of transistors and their interconnecting wires. Designers can usually simulate only some logic cells at this level in order to characterize them, but it is often impractical to simulate whole designs at this level.

It is important to note that a complete design can also be represented using a mixed approach; for example, a mixed model could be composed of a general structure at the RTL abstraction level, some other logic blocks described at behavioral level, and some blocks described at gate level.

Figure 5.1 represents the different levels of a typical design cycle, correlating the abstraction levels with some of the most popular languages used to describe the circuit in each phase. It must be noted that most common hardware description languages (HDLs), such as *VHDL* and *Verilog*, can be used at all top three levels of abstraction detailing in minor or major detail the circuit models.

Throughout the design cycle, design models could be produced resorting to very different approaches, including manual and automated ones. Regardless of the exploited approach, delivered models at every abstraction level are very likely to be inserted with *design errors* (i.e., design differences with respect to the desired specifications) that must be repaired.

Design debugging, i.e., identifying and removing design errors, is neither a trivial nor a cheap task; in fact, currently the budget devoted to perform this task approaches 60% of the total cost in the design process. [1]

To guarantee that the implemented circuit meets the desired specifications, every time a new model is delivered, an audit process must be performed to ensure that the delivered model complies with the initial specifications. It would not be appropriate, on the contrary, to perform a single *verification* step of the implemented circuit at the end of the whole manufacturing process: in fact, there is no doubt that the sooner a bug is detected the lower is the cost for correcting it. Uncovered bugs resulting from inadequate verification at early stages generate expensive situations, like the famous FDIV bug in the Pentium® processor. [2]

The terms *validation* and *verification* are quite often used by the research community and deserve some attention. The *Institute of Electrical and Electronics Engineers* in the *IEEE Standard 1012–1998* defines *validation* as, "The confirmation by examination and provisions of objective evidence that the particular requirements for a specific intended use are fulfilled;" and *verification* as, "The confirmation by examination and provisions of objective evidence that specified requirements have been fulfilled." [3] In the field of design, this definition is usually synthesized, stating that, "Design validation is concerned with building the right product, while design verification is concerned with building the product right."

Speaking about digital circuit design, when the physical circuit is not yet built (i.e., during early design stages), the demonstration that the intent of a design is preserved in its implementations has been called *design verification*. In other words, *design verification* is the process of verifying that all modeled behaviors of a design are consistent with another model. The reference model may represent a set of properties that the system needs to fulfill, such as, "All cache reads are consistent," but it is usually a higher-level design described at a different abstraction level.

In this chapter, we will call *verification* the process that aims at guaranteeing the correct translation of the model delivered at a certain abstraction level to its successive model. The reader should be aware that, despite their exact definitions, in the technical literature the terms *verification* and *validation* are sometimes used interchangeably.

If the circuit is physically built, the audit process performed on the actual circuit is called *test* [4] and should mainly aim at identifying either possible imperfections in the manufacturing process or defects introduced during the same process.

Usually, the model of the circuit to be verified, validated, or tested is called the *device under test* (DUT).

Design verification methodologies have been developed in a multi-flavored spectrum, ranging from *manual verification* techniques to *formal verification* techniques, and including, for example, random and semi-random approaches. Briefly, verification methodologies can be generically defined either as *formal* or *simulation-based*, and both methods exploit *dynamic* or *static* approaches, emphasizing the capacity to consider – or ignore – time information.

Formal verification methodologies use exhaustive mathematical techniques to prove that circuit responses to all possible inputs and all possible reachable states of the circuit conform to the specifications. These methods do not rely on the generation of input information to verify the design.

Generally speaking, formal verification may always be seen as theorem-proving within a given logic system. [5] However, in practice, research in this field falls within various sub-categories, such as: *automated theorem proving*, *model checking*, and *equivalence checking*. Since formal verification is out of the scope of this chapter, herein only a brief description is drafted:

- *Automated theorem proving*
  This is the oldest and most general form of formal verification, and it has been studied and practised since the 1960s. The idea is to represent two models or properties as two formulas f and g in a reasonably expressive logic, then prove $f \Rightarrow g$. While such an approach is potentially very general, results in automated theorem proving are limited algorithmically. Most implications in a general context are undecidable, and tackling them requires extensive user involvement. However, in restricted logics, such as *linear-time temporal logic* (LTL), proving the theorem $f \Rightarrow g$ is equivalent to testing the unsatisfiability of the formula $f \wedge \neg g$, which is always decidable. [6]

- *Model checking*
  This can be seen as a special case of proving the theorem $f \Rightarrow g$, where f is a state transition model and g is a logical formula stating a property or specification of $f \Rightarrow g$. The computational tree logic (CTL), introduced by Clarke and Emerson in 1982, [7] is commonly used, since it enables to carry out the check in a time linearly dependent on the size of f and g. Model checking is used, for example, to verify the control parts of a circuit, but it would be impractical to verify most circuit data paths using this method.

- *Equivalence checking*
  This is the most popular technique for formally verifying circuits. Essentially, approaches based on this method compare two models, and proceed in two phases: a Boolean network representing the design is extracted from the new model of the design (e.g., the model delivered at gate level); then, to verify the model, an equivalence evaluation is performed between the obtained network and a reference network, usually obtained at one of the previous steps of the design. The Boolean networks are usually represented using binary decision diagrams (BDDs). An introduction to BDDs used in formal verification is available in [8].

Equivalence checking is suitable for verifying two structurally similar designs that present a one-to-one mapping of circuit states. Since RTL models are usually produced by design engineers, gate-level models are the resulting process of several optimizations. Thus, state machines extracted from both models could be very different, since gate-level optimizations occasionally compact the circuit by merging some internal states or even by moving logic blocks from one side of a register to another. In fact, this method does not work as well comparing an RTL design with a gate-level one.

Computational resources (in particular, CPU time and memory) required to verify a design formally become significant even for medium complexity circuits.

On the other hand, when applicable, these methods provide valuable results, which are characterized by their exactness and independence of any specific input stimuli.

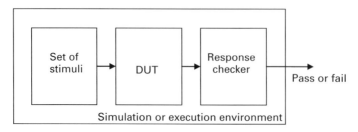

Figure 5.2   Audit environment

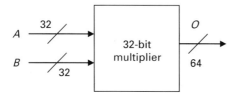

Figure 5.3   32-bit multiplier

Simulation-based methodologies aim at uncovering design errors by thoroughly exercising the current model of the circuit. Briefly speaking, a verification process based on circuit simulation requires three basic elements to be performed: input information (also called the *set of stimuli*), the model of the device under evaluation (also called the *design* or *device under test*), and finally the *response checker*, which generates the pass or fail information regarding the inspection process based on the comparison between the observed and expected behavior. It is clear that depending on the design stage, the audit process could be performed in different ways. For example, depending on the design state, the method could be based on a logic simulator or resort directly to the circuit (if the device has been built already).

Figure 5.2 shows the typical environment used to perform the audit process. This framework is valid if the proof has to be performed by simulating the circuit as well as if it is executed on the real circuit.

A *set of stimuli* is usually defined as the collection of inputs to the design under test or under verification. It could include, for example: configuration information, test patterns, instructions, and communication frames containing protocol errors to excite correction mechanisms. [8] On the other hand, the capacity of the set of stimuli to find bugs, errors, or faults in the device under test is called its *test quality*. In manufacturing testing, sets of stimuli are usually called *test sets*; however, herein the term *test sets* will be avoided.

Example 5.1 Let us analyze the combinational 32-bit multiplier shown in Fig. 5.3. The multiplier circuit has two 32-bit inputs *A* and *B*, and a 64-bit output *O*.

Exhaustively verifying the multiplier model by simulation using the verification scheme presented in Fig. 5.2 requires, as described already, the set of stimuli, the model under verification, and a response checker. Taking into consideration that the circuit is a combinational block, the set of stimuli that exhaustively verify the circuit must include all possible combinations of A and B ($2^{32} \times 2^{32}$).

Supposing that the multiplier is described in a hardware description language at RTL, and that the simulation of a single input configuration takes 1 μs (including the comparison process of the circuit outputs O with the attended values) the whole verification process will take about 584 thousand years of simulation.

In the 32-bit multiplier case, the search space of inputs includes 2^{64} possibilities. However, if the circuit is sequential, the complete space is further enlarged, owing to the number of states reachable by the circuit. In fact, if the circuit is sequential, it has to be activated with every possible input stimulus in every possible state. Thus, a sequential circuit counting on m primary inputs, and n state elements reaching, at most, 2^n states, must be exhaustively verified by exercising $2^m \times 2^n$ possibilities. It must be noted that while not all 2^n states may be reachable, the effort required to bring the circuit in a particular state might be not insignificant.

Based on the above examples, the reader can easily understand why simulation-based verification is rarely exhaustive: more frequently, the set of stimuli includes a carefully selected subset of the possible input configurations only.

Design engineers are usually interested in peculiar stimuli that deserve their attention. Some *special cases* may excite design particularities. Other stimuli, instead, may represent interesting cases that activate design functionalities in a singular manner. These stimuli are usually called *corner cases*.

In the case of the 32-bit multiplier, a set of special cases may contain some pairs of values containing, for example, one or both of the inputs A or B set to zero, or the multiplication of the largest values as well as the lowest ones, etc.

Tackling a more complex design, a set of corner cases built to verify a microprocessor core could include the execution of valid but unexpected instructions during the execution flow of an assembly program. For example, the execution of the ret instruction, which returns the program control from a subroutine, without the preliminary execution of the corresponding call instruction that invokes the subroutine is a corner case.

Special cases and corner cases should be included in a good set of stimuli, since they normally have a high error detection capability.

Simulation-based methodologies are strongly dependent on the quality of the set of stimuli used to excite the design. These methodologies are seldom exhaustive and only consider a limited subset of possible circuit behaviors. Since the quality of the results first depends on the percentage of applied stimuli with respect to the total number of possible stimuli, they almost never achieve 100% confidence of correctness.

It is obvious that different sets of stimuli could have different error detection capabilities, even if their length is comparable. Thus, regardless of the generation method, the key question is, how good is a set of stimuli produced at every design step?

Asserting the quality of a set of stimuli is, therefore, a major issue in the verification or validation area, which requires the introduction of new concepts. In fact, it is necessary to find appropriate mechanisms to produce sets of stimuli efficiently at every design phase.

A *test criterion* is a condition that defines what constitutes an *adequate* test. [10] Consequently, with respect to verification methodologies, the main idea consists of defining some testing criteria to assess the stimuli generation adequately. Each abstraction level in the design flow could count on appropriate metrics, strictly related to the description of the model, and be able to point out when the set of stimuli is satisfactory. Furthermore, choosing an adequate evaluation mechanism could help guide the verification process and determine when it can be terminated.

Coverage metrics were first defined in software testing as the measure of how thoroughly exercised a program is by quantifying the capacity of a given input stimulus to activate specific properties of the program code. [11]

Thus, borrowing the idea from software testing, a *coverage metric* for hardware verification can be defined to assure the adequacy of the set of stimuli, and the collected information about coverage could be exploited as a useful test criterion.

5.2   Coverage metrics

Taking into account the ideas sketched before, coupling stimuli-generation methods based on simulation with coverage analysis will provide information on how thoroughly a design has been exercised, driving the stimuli generation process without requiring any redundant effort. Coverage metrics can be used to guide the generation of sets of stimuli (e.g., as test criteria), as well as to evaluate the effectiveness of pseudo-random strategies by acting as heuristic measures that quantify verification completeness; finally, coverage analysis can help to identify inadequately exercised design aspects. [12]

Coverage metrics measure how thoroughly a design model has been covered (i.e., exercised) by a specific set of stimuli; usually, these measurements are expressed as a percentage value. [9]

The goal for a specific coverage metric could be implicitly or explicitly described with respect to the set of stimuli that is able to maximize the metric. For example, the *statement coverage* measures whether every statement of the source code is executed and it could be represented by a value in the range between 0 and 100%. It is clear that the metric does not provide any explicit information about the appropriate set of stimuli capable of getting the most out of this metric. On the other hand, a *functional metric* that evaluates whether a series of well-defined corner cases have been exercised presents explicit information about the set of stimuli in charge of maximizing the coverage. This metric could be also expressed as a percentage. However, it could be better to describe the obtained results using a check table.

The higher the coverage values obtained by a given set of stimuli, the higher the confidence in the design it can provide. Intuitively, high coverage values imply high system activation. However, it is worth noting that coverage does not imply that the design conforms to the specifications. A complete coverage on any particular metric could never guarantee a 100% flaw-free design, nor thorough code inspection. Moreover, it is not possible to select one single coverage metric as the most reliable. [12,13] Therefore, the current verification trend is to combine multiple coverage metrics, to obtain better results.

Performing a verification process guided by coverage metrics allows the achievement of high design verification but limits the redundant efforts. Coverage metrics act as heuristic measures for quantifying the verification completeness and identifying inadequately exercised design aspects.

In 1974, Brian Kernighan, the creator of the C language, stated that, "Everyone knows debugging is twice as hard as writing a program in the first place." This statement concerns software design; however, the same implication can be stated in the case of circuit design. Thus, it is really important to define clearly the verification strategy to be adopted.

As mentioned by Piziali in [14], the real success of a simulation-based verification process relies on the adequacy of the initial verification route-map, called the *functional verification plan*. A verification plan must define important test cases targeting specific functions of the design, and it must also describe a specific set of stimuli to apply to the design model. The verification plan can also allocate specific validation tasks to specialized engineers. Roughly speaking, the verification plan is composed of three aspects:

- *Coverage measurement*, defining the verification problem, the different metrics to be used, and the verification progress;
- *Stimulus generation*, providing the required stimuli to exercise thoroughly the device adhering to the directives given;
- *Response checking*, describing how to demonstrate that the behavior of the device conforms to the specifications.

Usually, the coverage metrics included in the verification plan are selected based on the designers' expertise, the ease of applicability, and a strict evaluation of the computational resources. Depending on the abstraction level of the available DUT model, some coverage data could be irrelevant; therefore, coverage data must be collected as soon as it is practical but not sooner. It is important to collect adequate information: if there are some metrics that do not deserve attention, it is better to avoid spending time on collecting the related information. Collecting meaningless information does not mean performing a real verification progress.

When several coverage metrics are used, it might be better to combine the collected values into a single overall coverage metric, using, for example, a weighted average of several coverage metrics. Let us consider three different metrics, CM_1, CM_2, and CM_3; a scale factor can be assigned to each coverage value, weighting properly the considered metric:

$$C_t = 0.3CM_1 + 0.5CM_2 + 0.2CM_3. \tag{5.1}$$

In this case, CM_2 is the most relevant metric of the set. Weighting metrics must be carefully performed, so as to both avoid masking small but essential contributions, and magnify meaningless information regarding any minor metric.

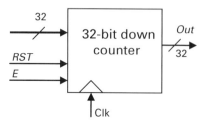

Figure 5.4   32-bit down counter

Unfortunately, no well-defined standard set of metrics able to guarantee acceptable design error detection exists. In fact, choosing a set of metrics to be included in the verification plan is a particularly critical task, which is customarily based on the experience acquired by the verification team.

Some issues should be taken into consideration when choosing coverage metrics. Specifically, it is important to:

• Evaluate the computational effort required to measure each coverage metric during the simulation process,
• Be able to generate and measure the test stimuli progress avoiding excessive efforts,
• Minimize the required modifications to the verification tools.

Additionally, it seems that no metric is superior to another; then a comparative measure of metric goodness needs to be established by intuitive or empiric experience. [10]

As mentioned before, an exhaustive circuit verification tries to evaluate thoroughly the whole set of possibilities of the stimuli–response space. However, even for a medium complexity circuit, the number of possibilities is quite large.

Aiming at reducing the space of stimuli and responses, it is possible to introduce a new concept: the *coverage model*. A coverage model is a subset of the stimuli–response space that will prove, with an acceptable degree of confidence, that the functionality of the design is correct. [9]

Example 5.2 Let us suppose that you need to verify the circuit model of the programmable 32-bit down counter shown in Fig. 5.4; each time it is reset, it loads a 32-bit word and the countdown starts. The counter exploits an additional control bit E to enable the counter operation. Therefore, the space of stimuli for this simple circuit consists of $2^{32} \times 2^2$ possibilities. Thus, assuming that verifying by simulating an input requires on average about 10 μs, about two days of simulation are required to verify the counter exhaustively. On the other hand, we could consider a coverage model that looks for special cases only. This means forcing the counter to load only a few values such as: all walking ones (32), all walking zeros (32), all ones (1), all zeros (1), and finally two particular cases: 55555555h (1) and AAAAAAAAh (1); for every configuration, the counter is forced to load it, and then the counter's ability to compute correctly the next value on the outputs is checked. Let us suppose also that for every input word, the counter is programmed twice: once to enable the countdown, and once to disable it. The results obtained from these experiments could still leave us with sufficient confidence about the correctness of the circuit model, and it is not necessary to spend a large amount of time to verify the design exhaustively, since the simulation of 136 input configurations only is required.

5.3   Classification of coverage metrics

In the following, some of the most common coverage metrics proposed in the literature and adopted in practice will be described and discussed. An introduction about the basic metric concepts is provided and in some cases specific examples are included. The classification presented here mainly follows what is proposed in [12].

5.3.1   Code coverage metrics

Code coverage metrics directly derive from metrics used in software testing. These metrics identify which code structures belonging to the circuit description are exercised by the set of stimuli, and whether the control flow graph corresponding to the code description has been thoroughly traversed. The structures exploited by code coverage metrics range from a single line of code to *if-then-else* constructs.

5.3.1.1   Statement coverage (SC)

Statement coverage is the most basic form of code coverage: statement coverage is a measure of the number of executable statements within the model that have been exercised during the simulation. Executable statements are those that have a definite action during runtime and do not include comments, compile directives (or declar-ations), etc. Statement coverage counts the execution of each statement on a line individually, even if there are several statements on that line.

Example 5.3 Let us assume that it is necessary to verify the model of an ALU described in VHDL at RTL. The scheme of the ALU is illustrated in Fig. 5.5. The ALU has 2-bit input signals A and B.

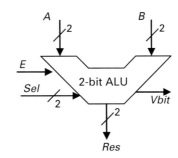

Sel
0 0 → Add numbers
0 1 → Subtract numbers
1 0 → Logic AND
1 1 → Logic OR

Figure 5.5   2-bit ALU

If the enable signal *E* holds the value 1, the ALU performs one out of four operations, otherwise the outputs become 0. The results of the performed operations are shown in the 2-bit signal and eventually in *Vbit*, the overflow bit.

In the VHDL description of the circuit presented in the following, the executable statements of the code have been enumerated to facilitate analysis.

```
entity ALU is
port( A:   in stdlogicvector(1 downto 0);
      B:   in stdlogicvector(1 downto 0);
    SEL:   in stdlogicvector(1 downto 0);
    E:   in stdlogic;
    RES:   buffer stdlogicvector (1 downto 0);
    Vbit:   out stdlogic
);
end ALU;
architecture RTL of ALU is
begin
  process(A,B,SEL,E)
  begin
1  RES 00;
2  Vbit 0;
3  if (E 1) then
4    if (SEL 00) then
5      RES A B;
6      if (A(1) 1 and B(1) 1 and RES(1) 0)
       or (A(1) 0 and B(1) 0 and RES(1) 1) then
7        Vbit 1;
8      else
9        Vbit 0;
10   elsif (SEL 01) then
11     RES A-B;
12     if (A(1) 1 and B(1) 0 and RES(1) 0)
       or (A(1) 0 and B(1) 1 and RES(1) 1) then
13       Vbit 1;
14     else
15       Vbit 0;
16   elsif (SEL 10) then
17     RES A and B;
18   else
19     RES A or B;
     end if;
   end if;
  end process;
end RTL;
```

The ALU description consists of some *if* statements, which select the different operations to be performed. In each branch of the *if* statements, assignments are made to the output signals (res and *Vbit*) according to the value of the control signals (*E* and sel). It is interesting to note that lines 6–9 as well as 12–15 are devoted to computing the overflow.

A set of stimuli devised to verify the circuit description must be composed of some stimulus, which contains a complete input signal configuration for the ALU (i.e., valid values for both control and input signals). In the following each stimulus is described using this format: (*A*, *B*, sel, *E*).

$$ALU\text{-}Set1 = \{(00, 01, 00, 1), (11, 10, 00, 1), (01, 01, 01, 1),$$
$$(01, 11, 00, 1), (10, 01, 10, 1), (01, 00, 11, 1)\}.$$

ALU-Set1 maximizes the statement coverage of the 2-bit ALU by executing all code lines at least. The first two stimuli are devoted to exercising the ADD operation, the next two undergo the SUB operation and the last input vectors exercise the AND once, and the OR once, respectively.

5.3.1.2   Branch coverage (BC)

Branch coverage reports whether Boolean expressions tested in control structures (such as the if statement and the while statement) evaluate to both true and false. The entire Boolean expression is considered as one true-or-false predicate regardless of whether it contains logical AND or logical OR operators. Branch coverage is sometimes called decision coverage.

Considering the set of stimuli of Example 5.3, devised to maximize the statement coverage, it is possible to note that branch coverage is not saturated, since the statement 3 is never false; then, BC could be easily maximized by adding the following set of stimuli:

$$ALU\text{-}Set2 = \{(00, 01, 00, \mathbf{0})\}.$$

Example 5.4 Consider the following piece of code:

```
   . . .
   wait until RST='1';
   . . .
```

Suppose that the available set of stimuli does not contain any stimulus causing RST to assume value 0. Even if statement coverage could be maximized, the branch coverage cannot be 100% covered because RST='1' will always be true.

5.3.1.3   Condition coverage (CC)

Condition coverage can be considered as an extension of branch coverage: it reports the true or false outcome of each Boolean sub-expression contained in branch statements, separated by logical AND and logical OR if they occur. Condition coverage measures the sub-expressions independently of each other.

Recalling Example 5.3, *ALU-Set1* does not allow CC to reach 100% coverage because the Boolean sub-expressions of the statements 6 and 12 are not thoroughly exercised. The reader is invited to devise a set of stimuli able to maximize CC of the 2-bit ALU.

Example 5.5 Let us consider the following VHDL code fragment:

```
. . .
CONTROL: process (INPUT)
begin
      case INPUT is
            when (3  5)
                  Z  A and B;
            when 1
                  Z  A or B;
            when others
                  null;
      end case;
end process CONTROL;
```

Let us assume that there are two available sets of stimuli for the control variable INPUT ranging from 0 to 7:

$$set_1 = \{(0), (1), (3)\},$$
$$set_2 = \{(0), (1), (3), (5)\}.$$

In both cases, the branch coverage equals 100%; however, set_2 reaches a higher value in the condition coverage, since this set of stimuli thoroughly excites the "when $(3 \mid 5) =>$" statement.

5.3.1.4   Expression coverage (EC)

Expression coverage is the same as condition coverage, but instead of covering branch decisions it covers concurrent signal assignments. It builds a focused truth table based on the inputs to a signal assignment using the same technique as condition coverage.

Maximizing EC for Example 5.3 requires a set of stimuli larger than the sets exploited to maximize SC, BC, and CC, since signal res must be activated thoroughly.

5.3.1.5   Path coverage (PC)

A circuit description could be embodied by its *control flow graph* that schematizes, using a graph, all possible paths that may be traversed at simulation time. Path coverage refers to the number of exercised paths present in the control flow graph of the circuit during the simulation of a set of stimuli. The number of paths in a circuit description that contains, for example, a loop structure could easily be very large.

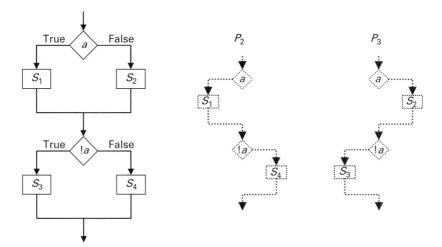

Figure 5.6   Control flow graph

Then, it is not feasible trying to cover all paths in a circuit. Instead, a representative set of reduced paths may be chosen.

Example 5.6 Consider the following VHDL fragment.

```
signal a : bit;
begin
    if (a) then
        { - sequence 1 of statements
    } else
        { - sequence 2 of statements
    }
    end if;
    if (!a) then
        { - sequence 3 of statements
    } else
        { - sequence 4 of statements
    }
    end if;
end;
```

Figure 5.6 shows a control flow graph representation of the piece of code previously presented. Control statements such as "if (a) then" are represented by decision points, while statement sequences 1, 2, 3, and 4 are depicted as the elaboration squares S_1, S_2, S_3, and S_4.

In the control flow graph in Fig. 5.6 it is possible to identify four possible paths: $P_1 = (S_1, S_3)$, $P_2 = (S_1, S_4)$, $P_3 = (S_2, S_3)$, and $P_4 = (S_2, S_4)$. However, taking into consideration the actual information on the control statements, it is only possible to traverse half of them (P_2 and P_3).

Today, most simulation and verification tools allow easy measurement of code coverage metrics. However, we already mentioned that a thorough coverage in code metrics is not enough to guarantee complete circuit verification.

Supposing that a collected value Ct of coverage must be computed to gather the contributions of all code coverage metrics obtained when applying the set of stimuli *ALU-SET1* to the ALU design described in Example 5.3, and assuming the following scale factors for each coverage metric: 0.2 for SC, 0.2 for BC, 0.3 CC, 0.2 EC, and 0.1 PC; the reader is invited to compute Ct.

5.3.2   Metrics based on circuit activity

Metrics based on circuit activity measure the activity of some portions of the design. These metrics mainly target interconnections, memory elements, and internal networks. Actual details about the final circuit structure may not be present at higher abstraction levels of the design. In fact, valid structural information at these levels of design is only present in module interconnections. At lower levels, however, memory elements and circuit networks come near to the actual circuit description; therefore, coverage metrics targeting activation of circuit structures are better suited at lower levels. However, as mentioned before, models at lower abstraction levels contain huge quantities of information that make extensive simulations unaffordable. In the following, the measure of the toggle activity will be considered as an example of these metrics.

5.3.2.1   Toggle coverage (TC)

Toggle coverage reports the number of nodes or storage elements that toggle at least once from 0 to 1 and at least once from 1 to 0 during the execution of a program. At the RTL, registers are targeted and, since RTL registers correspond to memory elements with an acceptable degree of approximation, the toggle coverage is an objective measure of the activity of the design. Indeed, this is a very peculiar metric and can be sensibly used to guarantee high activity in circuits present after the targeted registers.

Example 5.7 A description at RTL of the 32-bit multiplier is presented in Fig. 5.7. Let us suppose that it is necessary to maximize the toggle coverage of the input and output registers of the circuit.

By applying the short sequence of input values contained in the following set, it is possible to reach 100% of TC (over the input and output registers).

$setTC = \{(\dots 00h, \dots 00h), (\dots AAh, \dots AAh), (\dots 55h, \dots 55h), (\dots 00h, \dots 00h),$
$(\dots 55h, 02h), (01h, \dots 55h), (\dots 00h, \dots 00h)\}.$

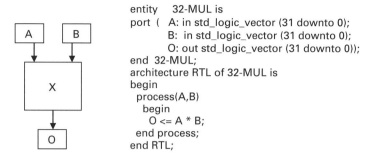

```
entity    32-MUL is
port (  A: in std_logic_vector (31 downto 0);
        B:  in std_logic_vector (31 downto 0);
        O: out std_logic_vector (31 downto 0));
end  32-MUL;
architecture RTL of 32-MUL is
begin
  process(A,B)
    begin
      O <= A * B;
   end process;
end RTL;
```

Figure 5.7   RTL description of a 32-bit multiplier

The value *setTC* is composed of seven pairs of values for the *A* and *B* inputs of the 32-bit multiplier; in the set, values preceded by "..." mean that the next hex value is repeated for all eight nibbles of the 32-bit word. The first four pairs guarantee the toggle coverage of the inputs, whereas the final three pairs act over the output.

5.3.3   Metrics based on finite-state machines

A *finite-state machine* (FSM) is a representation of a sequential circuit formed by modeling its behavior based on states, transitions, and actions. Thus, a coverage metric that represents the covered states of the circuit's FSM must be able to measure, quantify, and exercise the circuit sequential behavior. The classical representation of the finite-state machine of a digital circuit is a connected graph (normally called a state graph) corresponding to the states reachable by the circuit. Two metrics can be defined on this graph.

5.3.3.1   FSM transition coverage (FSM-TC)

Transition coverage measures the number of distinct edges traversed during the simulation of the set of stimuli.

5.3.3.2   FSM state coverage (FSM-SC)

State coverage measures the number of distinct nodes visited during the simulation of the set of stimuli.

Example 5.8 Let us consider the state graph of a 2-bit down counter (shown in Fig. 5.8). In this case, the circuit has a 2-bit control word (*ctr*) that allows the circuit to operate in one out of four operation modes: 00 *hold*, 01 *count*, 10 *load*, and 11 *reset*. The initial value is provided by a 2-bit input word (*in*). Whenever a new input value is loaded (*ctr* = 11), the down counter loads the 2-bit word and the countdown may start if the control word becomes 01. Finally, the counter holds the current value if the control word assumes value 00, and it is reset by applying 11 to *ctr*.

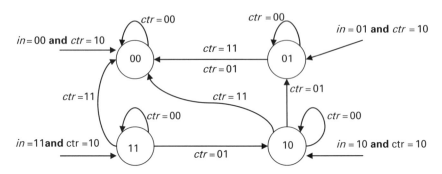

Figure 5.8   State graph of a 2-bit down counter

Let us suppose that it is necessary to generate a set of stimuli able to reach 100% coverage with respect to the FSM-SC metric. Clearly, a simple set of stimuli that provides the input word 11 is able to reach all possible states of the state graph by applying some clock cycles. On the other hand, assuming that FSM-TC must be maximized, this requires a set of stimuli larger than the previous one, since each edge in the state graph must be traversed. Indeed, to be able to traverse the edges related to the *reset* function of the counter ($ctr = 11$), it may be necessary to load each input word and then reset the circuit without counting.

Example 5.9 The following segment of code is a summary of the VHDL description of the control unit of an i8051 microprocessor. [15]

```
architecture bhv of controlunit is
type statetype is (reset,fetch,decode,execute,
incpc);
type executetype is (exe1, exe2, exe3, exe4, exe5);
signal cpustate: statetype;
signal executestate: executetype;
begin
  process(clock, rst, IRword)
    variable OPCODE: stdlogicvector(3 downto 0);
  begin
    if rst0 then
      . . . reset all              - Reset state
      cpustate reset;
      executestate exe1;
    elsif (clockevent and clock1) then
      case cpustate is
        when reset              - RESET pc
          PC 000000000000;
          - Reset memory
```

```
                   . . .
               cpustate fetch;
           when fetch          – Fetch instruction
             PCincPC 1;
             IRread 1;
                   . . .
               state decode;
           when decode          – Decode the instruction
             OPCODE : IRword(15 downto 12);
               case OPCODE is
                   when NOP =>=>
                     state incpc;
                   when MOV1 =>=>
                     case executestate is
                       when exe1
                            – MEM address;
                            executestate exe2;
                       when exe2
                            – Load memory value to register
                            executestate exe1;
                            cpustate incpc;
                   when MOV2
                       case executestate is
                           when exe1
                           . . .
                   when ACC_and_MEMDIR =>=>
                       case executestate is
                           when exe1
                           . . .
                   . . .
                   when others
                         state fetch;
                   end case;
               when incpc          – increment PC
                   . . .
               state fetch;
           when others null;
         end case;
       end if;
     end process;
   end bhv;
```

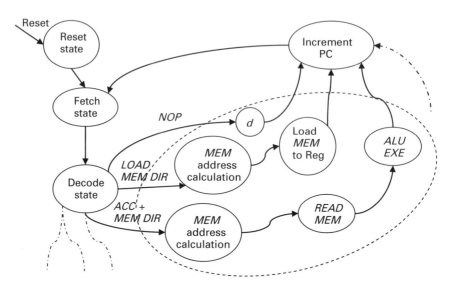

Figure 5.9   State graph for the i8051

The i8051 control unit can be modeled as an FSM. Figure 5.9 shows a state graph corresponding to the bolded lines of the previous piece of code. It is possible to identify clearly four states of the microprocessor: reset, fetch, decode, and increment PC. The system remains in these states for one clock cycle, only, while it remains in the execution state, represented by the dashed circle, or for a different number of clock cycles, depending on the instruction type to be executed. Figure 5.9 does not contain all the possible instructions of the microprocessor; however, three different instructions are considered. The *NOP* instruction requires only one clock cycle at the execution stage that actually does not perform any useful computation, while *LOAD MEM DIR* demands two clock cycles, and *ACC + MEM DIR* takes three clock cycles.

Assuming that the set of stimuli is composed of assembly programs, it is easy to reach 100% of coverage in the FSM state metric (FSM-SC) by executing at least once every type of instruction existing in the i8051 instruction set.

5.3.4   Functional coverage metrics

As the name suggests, functional coverage metrics target design functionalities during the verification process. These metrics are composed of a set of very precise test cases that exercise the design in well-defined situations, guaranteeing that the design under evaluation complies with some design functionalities. Often, the functions to be covered by these test cases are summarized in a check table.

In a verification team, the team manager is in charge of developing a group of tables containing the main functionalities to be tackled by the verification process. These tables are then distributed to skill engineers, who must create the appropriate sets of stimuli depending on the assigned tasks.

Table 5.1 Checkboard to verify the i8051 functionally

Test program description	Memory size (bytes)	Time (clock cycles)	Check box
All logic instructions	XXX	YYY	✓
All arithmetic instructions			
All memory instructions			
All jump instructions			
All input or output instructions			
All addressing modes			

Note that functional coverage deals with the functions implemented by a circuit, and not by its implementation: therefore, from the latter it is not possible to compute any functional coverage metric. On the other hand, it is worth noting that an advantage of these metrics is that they are independent (and can be computed independently) of the circuit implementation.

Example 5.10 Table 5.1 explicitly describes the initial set of test programs required to verify functionally the RTL description of the i8051 processor previously outlined.

The verification engineer must complete the table as soon as the results are available. In the proposed example in Table 5.1, the sets of programs to be performed elaborate different types of instructions. The table not only allows one to check whether all circuit functions have been addressed, but also to verify the cost (in terms of memory size and duration) of the corresponding test programs.

5.3.5 Error- (or fault-) based coverage metrics

Even though every one of the presented metrics here tries to exercise the circuit model finding errors present in the model of the circuit under evaluation better in some cases there is a loose relationship between the coverage metric and the actual design errors. On the contrary, error- and fault-based coverage metrics do not rely on the description format on which the model of the design is described, since these metrics are directly related to specific errors or fault models.

5.3.5.1 Mutation coverage (MC)

Design errors cannot easily be mapped to well-defined error models; however, in some cases, design errors due to typographical mistakes produced by the design engineer during the typing process have been successfully modeled by artificial *mutations* of the code.

An interesting way to measure how well design errors are likely to be detected or not by a set of stimuli consists of inserting some artificial errors into the design under verification, and then checking whether they are detected by the available set of

Table 5.2 Possible mutations of line 5 in Example 5.3

Original	Line 5 mutations
$Res <= A + B;$	$Res <= A / B;$
	$Res <= A * B;$
	$Res <= A - B;$

Table 5.3 Possible mutations of line 17 in Example 5.3

Original	Line 17 mutations
$Res <= A$ AND $B;$	$Res <= A$ OR $B;$
	$Res <= A$ XOR $B;$

stimuli. To calculate the test quality of the set of stimuli, the original model (without any change) is simulated in parallel with the model artificially changed, and then if the circuit outputs are different, the injected fault is detected. *Mutation coverage* is a typical example of coverage metric based on this approach.

This coverage metric mimics mutation coverage from software testing. In particular, this metric models a hypothetical design error by injecting a small local mutation in the design description. The resulting description, called a *mutant*, is simulated in parallel with the original design using the same set of stimuli.

Considering a fragment of code at the RTL, it is easy to inject small mutations into the design under verification. The following code mutations are a representative subset of possible design errors:

- Replacement of arithmetic operators,
- Changing constant values,
- Replacement of relational operators,
- Replacement of variables in operations and assignments,
- Replacement of logical operators,
- Deletion of operands from arithmetical operations.

Example 5.11 Considering the VHDL code fragment of the 2-bit ALU exposed in Example 5.3, and supposing that only some mutations of the arithmetic $(+, -, /, \times)$ and logic (AND, OR, XOR) operators are available, the possible mutations in lines 5 and 17 are shown in Tables 5.2 and 5.3.

The reader is invited to generate a set of stimuli able to maximize the mutation coverage of the code lines mutated as described in Tables 5.2 and 5.3.

5.3.5.2 Coverage metrics using fault models of manufacturing testing

Fault models try to bridge the gap between the physical reality of the circuits and the circuit-design description. Actually, regarding manufacturing testing, a fault model tries to model physical defects that may appear in the circuit after manufacturing. For example, sticking one of the input or output signals of a logic gate at a predetermined logic value (1 or 0). This fault model is called *single stuck-at* 0 or 1, and it is the most popular fault model used in digital circuit manufacturing tests. [4] Usually, to facilitate coverage measurements, it is assumed that only one fault is present in the circuit model at the simulation time. A set of stimuli covers a fault if the circuit containing that fault causes a different output behavior of the design with respect to the original behavior of the fault-free circuit.

Some fault models (e.g., single stuck-at) are easy to simulate using modern logic simulators, and there are tools (named automatic test pattern generators, or ATPGs), that are able to generate sets of stimuli tackling them. For these reasons, it is sometimes convenient to use some metric based on these fault models to evaluate the test quality of a set of stimuli devised for verification.

Coverage metrics used in manufacturing testing are strongly related to fault models emulating physical errors. These fault models are broadly used to assert the test quality of test sets produced for post-production testing. As mentioned before, the most popular fault model is the single stuck-at fault; however, the research community has defined several fault models that must be taken into consideration when performing a set of stimuli. Since a deep description of manufacturing fault models is out of the scope of this chapter, only the most commonly used fault models are listed here: stuck-at, bridge, delay, transition, and cross-talk.

5.3.6 Coverage metrics based on observability

Figure 5.2 schematizes a classical simulation-based design verification environment. Commonly, verification is performed by comparing circuit responses with a reference database obtained previously. However, it is also a common practice to verify the circuit design by running the DUT in parallel with a reference model described at a different level of abstraction (usually called a *golden model*) or by the addition of circuit monitors or internal assertions that confirm the correct behavior of the DUT. [6] Running the DUT and a golden model in parallel avoids the need to check circuit primary outputs each time. It could also be possible to *observe* some internal variables on the DUT by checking their behavior against the golden model variable's behavior. In some cases, it is possible to observe only a reduced set of variables; thus, the remaining variables are usually called *unobserved variables*.

Coverage metrics based on observability are able to determine a discrepancy from the desired behavior of the DUT by observing whether an observed variable takes a value that conflicts with the expected value specified by the reference model.

5.3.6.1 Observability-based code coverage metric (OCCOM)

The observability-based code coverage metric was introduced in [7]. In this approach, the DUT is considered as a structural interconnection of modules. The modules can be composed of combinational logic and registers. Given a set of stimuli, and using a logic simulator, controllability metrics can easily be computed. The authors of OCCOM define a single tag model. A *tag* at a code location represents the possibility that an incorrect value was computed at that location. For the *single tag* model, only one tag is identified and propagated at a time. The goal, given a stimulus and a description of the model, is to determine whether (or not) tags injected at each location are propagated to the primary outputs of the system. The percentage of propagated tags is defined as the code coverage under the proposed metrics. A two-phase approach is used to compute OCCOM: first, the circuit description is modified, eventually by the addition of new variables and statements, and the modified descriptions are simulated, using a standard simulator; second, tags (associated with logic gates, arithmetic operators, and conditions) are then injected and propagated, using a flow graph extracted from the circuit description.

In conclusion, in a coverage-driven verification process, the strategy becomes how to use the best implementation vehicle to hit most of the points with the least effort.

5.4   Coverage metrics and abstraction levels of design

Now that the different categories of metrics have been introduced, it is important to underline once more that the choice of the most suitable metric (or set of metrics) to achieve a given verification goal for a given design is a very critical task, which must be performed taking into account several parameters, including the level of abstraction of the currently available description of the circuit. In fact, the level of abstraction strongly affects the choice.

First of all, it is important to consider the ease of computation of each metric (e.g., in terms of simulation cost for measuring the metric itself); moreover, the cost depends on whether some additional or unavailable information is required to compute the metric. Finally, some thought has to be given to the current state of the art, looking for recognized metric efficiency in terms of obtained results.

Functional coverage, mutation coverage, and coverage metrics based on FSMs can be computed even when high-level descriptions, only, are available, and implementation details still have to be decided; functional and FSM metrics are mainly concerned with design functionalities, whereas mutation metrics focused on design errors are more likely to be used at high abstraction levels. Additionally, the cost required to measure these metrics at high abstraction levels is relatively low.

Verification processes performed at register level exploit code coverage metrics better, since this set of metrics has been developed to target specific software structures that are exploited very often at this abstraction level.

On the other hand, observability-oriented, toggle-based, and manufacturing metrics are more strictly related to low-level circuit implementation details. Indeed, coverage metrics targeting circuit activation of circuit structures are better suited at lower levels of abstraction.

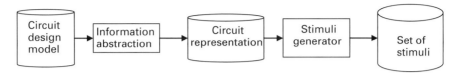

Figure 5.10 Open-loop generation scheme

5.5 Stimuli generation methods

While it is accepted that implementing a simulation or execution environment, as described in Fig. 5.2, is not an easy mission, probably the most difficult task is stimuli generation. The research community has deeply studied the stimuli generation domain, and the proposed methodologies range from manual generation to random approaches, including deterministic and heuristic methods. Even if the methods described herein target design verification, these have also been called test-generation methods.

A *stimulus* has been defined in the introduction as the input to the design under test or under verification. A set of stimuli might include, but is not limited to, configuration information, test patterns, transactions, instructions, assembly programs, exceptions, and data packets. [9] The kind of stimulus and its characteristics depend on the circuit under analysis and on the level of abstraction the analysis is performed on.

Depending on the control scheme exploited to generate stimuli, generation methodologies can be defined as either *open-loop* or *closed-loop*, depending on whether the method uses feedback information or not. While both methodologies directly depend on the circuit description, regardless of the abstraction level, open-loop methodologies generate stimuli resorting to the previously acquired test experience in terms of design knowledge as well as data gathered in former verification campaigns. The main difference between these methods is the use or not of feedback information during the stimuli-generation process.

Figure 5.10 represents the work flow used to generate a set of stimuli resorting to open loop mechanisms (also called *feedback-based*). As shown in Fig. 5.10, the first step in the open-loop scheme consists in the abstraction of valid information about the circuit. This internal representation of the circuit could be based on very simple *constraints* about the possible values to be assumed by some inputs, or even complex graphs representing all possible interconnections among memory elements in the design. Once the circuit representation is acquired, the set of stimuli is generated, resorting, for example, to some automatic algorithm or manual strategy.

In 1998, a program called VERTIS was introduced [18]; this algorithm is able to automatically generate assembly programs suitable for processor verification without using feedback information. VERTIS takes as input the assembly language instruction set of the processor, and the operations performed by the processor in response to each instruction, and produces an assembly program without exploiting feedback information.

The methodology is developed as follows: first, for each instruction, VERTIS is able to generate a sequence of instructions that enumerates all the combinations of the operations and systematically selects operands; second, a set of additional instructions

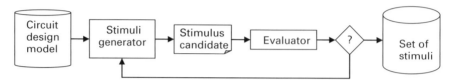

Figure 5.11   Closed-loop generation scheme

is added to the original program, and is able to compress a signature containing the valuable information about the execution of the program.

Closed-loop methodologies include in the generation processes an *evaluator* module that is able to elaborate the stimulus candidates and return a feed back value representing the stimulus quality. The basic scheme of the closed-loop methods is illustrated in Fig. 5.11; the stimuli generator creates a stimulus candidate that is evaluated, and then it is or is not added to the final set of stimuli depending on whether the new stimulus satisfies test quality conditions initially defined in the generation process set-up. For example, a logic-circuit simulator may be used as a stimuli evaluator for collecting code-coverage information regarding the simulation process of the device under verification.

The feedback information is usually exploited in one out of three possible aspects of the generation:

- *Threshold*
  The feedback information is used by the stimuli generator to determine when the set of stimuli is complete; otherwise, the stimuli generator will continue producing stimuli information. For example, the set of stimuli is considered complete when all code coverage metrics equal 90%; otherwise, the stimuli generator continues to produce new stimuli information.
- *Feedback-sifted generation*
  The feedback information is exploited to accept or discard new stimuli. For example, only new candidates able to increase the coverage in a set of selected metrics will be added to the final set.
- *Feedback-driven generation*
  The feedback information is used by the stimuli generator to indicate stimuli candidates that better comply with the quality conditions required in the process, and then to optimize them. Evolutionary algorithms usually exploit feedback-driven generation to produce sets of stimuli.

Stimuli generation could be performed following two mainly different methods: manual and automatic. Manual generation is performed by human beings, whereas automatic methods may be based on different approaches, such as random techniques and deterministic or heuristic algorithms.

5.5.1   Manual generation

Manual stimuli generation (also called *directed* generation) may be performed by verification, test, or even design engineers. Sets of stimuli are generated manually by following the verification strategy defined by the verification team. First, the verification strategy tackles the main functionalities of the design, and then, both special and corner cases are addressed for verification.

Regularly, the verification team employs expert engineers who have acquired a significant *know-how* while they were involved for example in circuit design. A verification engineer can manually write a set of stimuli tackling special and corner cases, which are really difficult to reach using other generation techniques.

For example, in the processor-design cycle, one of the most common methods used to verify the processor design initially consists of the generation of some assembly programs containing all the processor instructions. Afterwards, more sophisticated programs may be manually devised to excite specific design particularities.

To devise stimuli generation efficiently by hand, a deep knowledge about the device under evaluation is usually required. Let us suppose that a set of stimuli must be manually generated to verify the *handshaking* properties of a device originally devised to perform a communication protocol. It is clear that the verification engineer must be familiar with the handshaking properties of the communication protocol; moreover, some knowledge about the device implementation details might also be useful to generate the set of stimuli competently and more successfully look for possible design errors. In fact, sets of stimuli are really time consuming and expensive to generate by hand.

The main advantage of the manual generation of sets of stimuli is that this method does not require the development of complex and usually expensive tools to generate the stimuli automatically. Additionally, since this method mainly relies on the engineer's expertise, manual generation of sets of stimuli may start even sooner, and before detailed information about the device implementation is available.

On the contrary, the main disadvantage of manual generation techniques is that the resulting set of stimuli strongly relies on the engineer's knowledge and might be easily biased by his or her beliefs. Additionally, its cost can be very high, since it requires highly skilled engineers to generate effective sets of stimuli. Supposing that the engineer in charge of the verification process is the same design engineer; then, it is likely that a misconception introduced in the design step deriving a design error will not be detected by the set of stimuli generated.

5.5.2 Automatic generation

Manual approaches used to generate sets of stimuli do not provide a standard solution for the design-verification problem, because there is no guarantee of the accuracy of the results that can be achieved in terms of errors or bugs found. Thus, the research community has developed different techniques to generate sets of stimuli automatically, resorting to different approaches.

Methodologies to generate sets of stimuli automatically usually rely on software tools. Some of them resort to an internal representation of the device under verification, generating the sets of stimuli in two phases: first, a special representation of the design is extracted from the available model, in the case that this representation does not yet exist. This internal representation could be either that produced by the designer, or an ad-hoc one (possibly extracted from the latter), and based on graphs, state machines, or other representations. Second, the actual set of stimuli is generated exploiting the previously acquired information. On the contrary, some other automatic generation mechanisms directly use the design under verification to produce the sets of stimuli.

An automatic closed-loop method to generate a set of stimuli for digital circuit verification should be characterized by:

- High flexibility regarding the targeted circuit, to allow the maximum applicability of the method;
- Syntactically correct generation of input information depending on the specific singularities of the targeted design, for example, special constraints required in some input variables;
- High versatility with respect to the evaluation system, to allow the tackling of different verification problems at different abstraction levels;
- Ability to drive the generation process exploiting a feedback measure, such as a coverage metric.

In recent years, the research community has focused efforts on automatic generation of stimuli, trying to provide alternative techniques for the automatic generation of verification sets of stimuli. Delivered solutions have been mainly based on deterministic and random approaches; however, emerging methodologies based on different heuristics have demonstrated their suitability. Most of the deterministic approaches are open-loop-based strategies; conversely, several pseudo-random methods exploit closed-loop schemes.

Automatic generation of stimuli based on closed-loop mechanisms is really efficient; however, the main drawback regarding these methods is the computational effort involved to produce a good set of stimuli, because it is necessary to evaluate high quantities of candidates to complete the final set. Additionally, automatic generation requires an environment that in some cases is really difficult and expensive to set up.

5.5.2.1   Deterministic generation techniques

Deterministic approaches generate the stimuli-exploiting algorithms that use exact decision mechanisms, and mainly utilize the information extracted from the design under verification during the generation process.

Algorithms that automatically generate test vectors, aiming at testing through simulation, have been called ATPG (*automatic test-pattern generation*). The term ATPG was first introduced for manufacturing testing tools; however, this term is also currently accepted in design validation and verification. Some ATPG techniques can also be successfully exploited to generate sets of stimuli not only for testing, but for verification, too. It is worth noting that most of the ATPG-like algorithms exploited to generate automatically sets of stimuli maximize the coverage of the single stuck-at faults of the circuit.

The classical deterministic approach to generating test patterns suitable for manufacturing testing automatically was first described in [19]. This deterministic algorithm was called D-Algorithm (D-ALG) and it was the first widely accepted algorithm for automatic generation of test patterns. The D-ALG is capable of finding test vectors for all detectable faults in a given combinational circuit. The author developed a complete mathematical description in the D-ALG, based on the "calculus of the D-cube," which guarantees that a test will always be generated for detectable stuck-at faults.

Path-oriented decision making (*PODEM*) improved on the efficiency of the D-ALG by noting that all nodes in a combinational circuit are completely determined by the primary input values. [20] After identification of the target stuck-at fault, PODEM assigns input values one at a time and immediately simulates the results. This process guarantees that any conflict can be resolved by complementing the latest input assignment setting, otherwise, no solution can exist for the current set of input assignments. When conflict occurs for both logic states on the first input pin selected, the fault is guaranteed to be undetectable. The process starts by initializing all inputs to the unknown (X). Subsequently, the logic value of any given node within the circuit will either be X or will be dominated by an already selected logic value (from a previous input logic assignment). The search continues as long as no node value conflicts with the fault sensitization and propagation requirements. As the input logic assignment and simulation progress, any conflict is a function of the latest input logic assignment coupled with all current input pin logic assignments, since all X-valued nodes are dominated by the most recent logic selections. All combinations of input logic values for the yet-undetermined input pins need not be attempted because they cannot resolve the present conflict. A significant portion of the search space can thus be discarded without requiring computer time to search that portion of the space explicitly.

Tackling verification of combinational circuits, some algorithms based on PODEM have been developed to generate sets of stimuli automatically; for example in [21], the authors describe a verification algorithm, called *PLOVER-PODEM*, whose enumeration phase is based on PODEM. Parallel-logic verification schemes have been exploited.

Let us consider a sequential circuit design that must be verified. The verification process could follow the two-step strategy described before: in the first step, an FSM of the circuit is extracted from the available description of the design. Then, as a second step, a deterministic algorithm is exploited to generate stimuli able to traverse all edges of the FSM automatically.

Tackling microprocessor cores verification, in [22] the authors described an algorithm to extract a very small FSM that encapsulates the processor control behavior. The generation process is evolved in three phases: first, the FSM model is directly extracted from the design model. Second, all possible transition paths with a given finite length are generated; path coverage is used to measure the quality of the generated stimuli. Third, all the FSM state transition paths are translated to instruction sequences.

5.5.2.2 Random generation techniques

Random strategies are some of the most popular methods of generating sets of stimuli because they are simple and require low human intervention; these strategies randomly explore the stimuli search space, looking for acceptable information to complete the set of stimuli; in synthesis, random approaches introduce probabilistic decision mechanisms in the algorithm of generation.

Guaranteeing a thorough verification of the circuit is necessary to generate huge sets of inputs that could require massive computational resources in terms of memory and time. It is important to note that stimuli generation based on random approaches requires little effort in generating sets of stimuli automatically; however, most of the effort is consumed by the evaluation mechanism exploited to assert the stimuli quality. In fact, even statistical analyses have been proposed to decrease as much as possible the generation of redundant stimuli. [23]

Habitually, random strategies are equipped with initialization mechanisms that allow repeatability in the stimuli generation campaigns; for instance, most of the random generation tools are provided with the possibility of selecting the initial seed of the random function to produce repeatable results. Even though this fact inhibits the actual randomness properties of a stimuli generator, herein the term *random* is used anyhow (a more correct term would be pseudo-random).

Random strategies must be carefully configured before use because these techniques could generate large quantities of information incapable of improving the set of stimuli. In fact, some designs require specific sequences of inputs to be thoroughly verified; the kinds of circuit that are difficult to test or verify using random strategies have been called *random resistant* circuits, and in these situations random-based generation strategies could fail in reaching the target quality.

A well known random resistant circuit is a pipelined processor; indeed, verifying the correct design of pipelined microprocessors requires sequences of suitable instructions (e.g., to excite specific control mechanisms such as *data forwarding* [24]) that can hardly be generated randomly.

In a verification process based on random generation, it is very important to identify stimuli that can be created without any constraints from others that require constraints to ensure that valid stimuli are being created. Other stimuli require additional or modified constraints to create certain special cases.

The ability to create well-defined patterns between different input signals is directly related to the ability to express the constraints able to cause the pattern to be generated.

Let us suppose that it is necessary to generate data communication frames to verify a serial communication device using a random-stimuli generator; it is possible that this process becomes an inefficient task if special constraints correlating the *parity bit* with the *data word* if the internal structures of the data communication frames are not clearly defined.

As another example, let us consider a random constrained generator of a set of stimuli for processors called *Genesys*, which is a random-based generator of test programs suitable for microprocessor verification, developed by IBM. [25] The system consists of three basic interacting components: a generic, architecture-independent test generator, an external specification, which holds a formal description of the targeted processor, and a behavioral simulator used to predict the results of instruction execution. Additionally, the user can control the generation process by specifying desired biasing towards special events. As exemplified by the authors of [25], the result of zero for an ADD instruction is typically of special importance, whereas its relative probability of occurring randomly is practically non-existent, and should, thus, be generated with a reasonable probability.

Random techniques can be included either in an open-loop method, or in a feedback-based one. If included in a closed-loop strategy, the random generation mechanism will generate stimulus information until a stop condition is reached. Indeed, feedback-based strategies including random generation are usually merely cumulative: feedback information is not used to optimize stimuli, but a new stimulus is added to the test or verification set if it increases the test or verification quality of the whole set of stimuli, only. Therefore, at the end of the generation process, all the gathered information will compose the set of stimuli, regardless of the real value of the generated stimulus.

The main advantage of random strategies is the low human intervention required to perform the generation experiments. Additionally, it is interesting to highlight the fact that generation is not biased by the human understanding of the design behavior.

5.5.2.3 Emerging generation techniques

Emerging methodologies based on new heuristics are being successfully exploited to generate optimal stimuli automatically when it is impractical to find perfect solutions, owing to the high amounts of resources required in terms of computational time and memory. Evolutionary algorithms are sketched here as an interesting example of these kinds of generation mechanism.

Evolutionary algorithms (EA) are based on pseudo-random generation and are able to guide the generation process using feedback information. These optimization algorithms are based on the natural-selection paradigm.

Roughly speaking, in evolutionary computation there is an initial population of individuals. These individuals are evaluated using a fitness function, and the fittest ones (called *parents*) survive and produce new individuals (or *children*). These new individuals join the population and older individuals may be removed from it. The whole process is performed in genetic cycles called *generations*. New individuals are produced by either mutation or cross-over operators. Mutation operators make a random change to the parent; whereas cross-over operators perform recombinations of the parents.

Evolutionary algorithms have been successfully exploited to generate sets of stimuli for digital circuits described at RTL and gate level, [26] [27] and also for verifying complex circuits such as processor cores. [28]

5.6   Acknowledgements

The authors wish to thank Massimiliano Schillaci and Michelangelo Grosso for their valuable comments.

5.7   References

[1] Semiconductor Industry Association (2002). *International Technology Roadmap for Semiconductors 2002 Update*. www.semichips.org/pre_stat.cfm?ID=153.

[2] T. R. Halfhill (1995). The truth behind the Pentium bug. *Byte* (March). www.byte.com/art/9503/sec13/art1.htm.

[3] *IEEE Standard for Software Verification and Validation*, IEEE Std 1012–1998 (1998).

[4] M. L. Bushnell and V. Agrawal (2000). *Essentials of Electronic Testing for Digital, Memory and Mixed-Signal VLSI Circuits* Kluwer Academic.

[5] R. Kurshan (1994). *Computer-Aided Verification of Coordinating Processes*. Princeton University Press.

[6] W. W. Bledsoe and D. W. Loveland, eds. (1984). *Automated Theorem Proving: After 25 Years*, Contemporary Math 29, American Mathematical Society.

[7] E. M. Clarke and E. A. Emerson (1982). Design and synthesis of synchronization skeletons for branching time temporal logic. In *Proceedings Logic of Programs Workshop, LNCS* **131**, pp. 52–71. Springer-Verlag.

[8] A. J. Hu (1997). Formal hardware verification with BDDs: an introduction. In *IEEE Pacific Rim Conference on Communications, Computers and Signal Processing*, vol. 2, pp. 677–682.

[9] J. Bergeron, E. Cerny, A. Hunter, and A. Nightingale, (2006). *Verification Methodology for SystemVerilog*. Springer.

[10] H. Zhu, P. A. V. Hall, and J. May (1997). Software unit test coverage and adequacy. *ACM Computing Surveys*, **29**(4):366–427.

[11] J. B. Goodenough and S. L. Gerhart (1977). Toward a theory of testing: data selection criteria. In R. T. Yeh, ed., *Current Trends in Programming Methodology*, vol. 2, pp. 44–79. Prentice-Hall, Englewood Cliffs.

[12] S. Tasiran and K. Keutzer (2001). Coverage metrics for functional validation of hardware designs. *IEEE Design and Test of Computers*, **18**(4):36–45.

[13] J. L. Chien-Nan, C. Chen-Yi, J. Jing-Yang, L. Ming-Chih, and J. Hsing-Ming (2000). A novel approach for functional coverage measurement in HDL circuits and systems. In *IS-CAS2000: The 2000 IEEE International Symposium on Circuits and Systems*, pp. 217–220.

[14] A. Piziali (2004). *Functional Verification Coverage Measurements and Analysis*. Kluwer Academic Publishers.

[15] J. Simsic and S. Teran (2001). *8051 Core: Overview*. http://www.opencores.org/projects.cgi/web/8051/overview.

[16] M. Kantrowitz and L. M. Noack (1996). I'm done simulating; now what? Verification coverage analysis and correctness checking of the DECchip 21164 alpha microprocessor. In *The 33rd Design Automation Conference*, pp. 325–330. ACM Press.

[17] F. Fallah, S. Devadas, and K. Keutzer (2001). OCCOM: efficient computation of observability-based coverage metrics for functional verification. *IEEE Transactions on Computer-Aided Design*, **20**(8):1003–1015.

[18] J. Shen and J. A. Abraham (1998). Native mode functional test generation for processors with applications to self test and design validation. In *IEEE International Test Conference*, pp. 990–999.

[19] J. P. Roth (1996). Diagnosis of automata failures: a calculus and a method. *IBM Journal of Research and Development*, **10**:278–291.

[20] P. Goel (1981). An implicit enumeration algorithm to generate tests for combinational logic circuits. *IEEE Transactions on Computers*, **C-30**(3):215–222.

[21] H.-K. T. Ma, S. Devadas, R.-S. Wei, and A. Sangiovanni-Vincentelli (1989). Logic verification algorithms and their parallel implementation. *IEEE Transactions on Computer-Aided Design of Integrated Circuits and Systems*, **8**(2):181–189.

[22] J. Shen and J. A. Abraham (1999). Verification of processor microarchitectures. *IEEE VLSI Test Symposium*, pp. 189–194.

[23] Y. Malka and A. Ziv (1998). Design reliability estimation through statistical analysis of bug discovery data. In *35th Design Automation Conference*, pp. 644–649. ACM Press.

[24] D. A. Patterson and J. L. Hennessy (1996). *Computer Architecture – A Quantitative Approach*. 2nd edn. Morgan Kaufmann.

[25] L. Fournier, Y. Arbetman, and M. Levinger (1999). Functional verification methodology for microprocessors using the Genesys test-program generator. Application to the x86 microprocessors family. In *IEEE Design, Automation and Test in Europe Conference and Exhibition*, pp. 434–441.

[26] V. Hahanov, A. Babich, A. Sokolov, and V. Pudov (2002). Deterministic method of genetic algorithms of test generation for digital systems verification. In *Modern Problems of Radio Engineering, Telecommunications and Computer Science, IEEE Proceedings of the International Conference*, pp. 257–258.

[27] Z. Stamenkovic, H. Dahmen, and U. Glaeser (2001). VHDL design validation by genetic manipulation techniques. In *IEEE 22nd International Conference on Microelectronics*, vol. 2, pp. 735–738.

[28] F. Corno, E. Sanchez, M. Sonza Reorda, and G. Squillero (2004). Automatic test program generation – a case study. *IEEE Design and Test*, **21**(2):102–109.

6 SystemVerilog and Vera in a verification flow

Shireesh Verma and Ian G. Harris

6.1 Introduction

The goal of this chapter is to illustrate the practical applicability of the simulation-based validation concepts in the book by applying them to a design example. We will use both SystemVerilog [1] and Vera [2] as hardware-verification languages (HVLs) in which we will implement the entire validation framework for the design example. Simulation is the most widely used technique for verification of design models. The design to be verified is described in a hardware-description language (HDL) and is referred to as the design under verification (DUV). This provides an executable model or models of the DUV. These models could be developed at different levels of abstraction.

A high-level design specification is then analyzed to produce stimulus or input test vectors. The input test vectors are applied to the models. The inputs are propagated through the model by a simulator and finally the outputs are generated. A monitor is used to check the output of the DUV against expected outputs for each input test vector. It is constructed based on an interpretation of the expected design behavior from the specification. If there is any observed deviation from the expected output, a design error is considered to have been found, and debugging tools are used to trace back and diagnose the source of the problem. The problem usually arises from either incorrectly modeled design or incorrectly modeled timing. Once the problem source is identified, it is fixed and the new model is simulated. In an ideal world, the model should be tested for all possible scenarios. However, this would amount to generating an infeasible number of test vectors. Since only a limited number of test vectors can be simulated, the goal should be to identify and pick the most useful ones. The usefulness of a test case is usually defined by the extent to which it covers, or rather uncovers, the features of the design. Moreover, a test case that verifies an already verified part of the design does not add any value. Coverage metrics provide a measure of the degree to which a design has been verified.

Therefore, several coverage metrics have been invented to quantify the usefulness of a test case [3]. The simulation performance can be improved either by speeding up the simulator or by choosing test cases intelligently to maximize coverage with minimal

Practical Design Verification, eds. Dhiraj K. Pradhan and Ian G. Harris. Published by Cambridge University Press. © Cambridge University Press 2009.

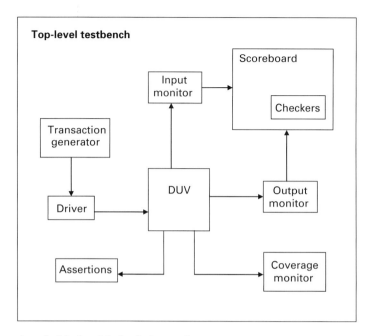

Figure 6.1 A typical industrial simulation environment

simulation runs. One optimization is to reduce test generation time by giving constraints to stimuli and testing with only valid inputs. Monitoring non-primary output variables in the model reduces debugging time by pointing out the error closer to its source. The success of this method of verification depends on both the quantity and the quality of the test vectors.

6.2   Testbench components

A typical industrial verification environment consists of a multitude of components. [4] They include the DUV, stimulus, monitor, checking, and scoreboard components. The complete assembly of such a set-up is called a testbench. Figure 6.1 shows such an environment. Different tool vendors might show additional or fewer components depending on how they choose to organize their tool flow. Figure 6.1 represents all the core functional aspects of a verification environment. In general, a testbench includes all the code needed to create, observe, and check a deterministic input sequence given to the design. This input sequence may be generated by a direct approach or by a random method. The testbench is a closed system in the sense that the top level of the test bench has no inputs or outputs. It is effectively a model of the universe from the standpoint of the DUV. In terms of work division, a verification engineer develops code for components of the testbench universe and the designer develops HDL description of the DUV. The former is typically written in a hardware-verification language (HVL), such as SystemVerilog or Vera, or a general-purpose programming language, such as C or C++, whereas the latter is developed using HDLs. The components developed in HVL or C or C++ communicate with the simulation engine through an API. The effective design of a testbench entails making sure that the design is stimulated with interesting input patterns, which cover as much functionality as possible and that the expected responses are computed based on these input patterns. The design can be ascertained to be functioning as intended by exercising all the functionality and by predicting and checking all responses. In the following sections, we present a brief discussion of each component of the verification environment.

6.2.1 Design under verification

The DUV sits at the core of the verification environment. Almost all of the other components interact with the DUV. The task of verification is to find the design errors present in the DUV. The DUV is represented by an HDL description of the design. During simulation, the HDL description is interpreted or compiled into a model capable of being simulated. The exact nature of the model depends on the simulator being used. The difference in these models arises from different kinds of interpreters and compilers used in different simulators.

The DUV can represent any level of hierarchy in the design. It may represent a macro, a logical unit, a chip, or an entire system. Irrespective of the level of hierarchy to which the DUV belongs, all the other components of the testbench are customized to exercise and verify it. The DUV may be described at any level of abstraction including behavioral, RTL, gate, or transistor level. Irrespective of the abstraction level, verification involves making sure that the DUV functionality matches its intent.

The stimulus components manipulate the inputs of the DUV and the checker and monitor components observe its outputs. Depending on the verification needs, monitor or checker components may be embedded inside the DUV in some cases.

6.2.2 Monitor

A monitor is a model that observes different aspects of a verification environment. It is a self-contained component that observes the outputs of the DUV for protocol correctness, the inputs and non-primary outputs of the DUV for functional coverage analysis and scoreboard updates, and the internals of the DUV for events of interest to the environment. A monitor cannot cause any side effect on the verification environment, since it does not drive any signal into the DUV; it only receives the inputs and callbacks to itself. So, it is reusable at other levels.

6.2.2.1 Input monitor

An input monitor collects the input signals to the DUV and passes it to the scoreboard which uses those data to compute expected outputs and compare them to the ones provided by the output monitor. The input monitors may have a checker component embedded in them, depending on the implementation.

6.2.2.2   Output monitor

An output monitor observes the outputs of the DUV. These outputs are compared with the expected outputs posted on the scoreboard and any mismatch is considered an error. The output monitors may have a checker component embedded in them, depending on the implementation.

6.2.2.3   Coverage monitor

A coverage monitor collects input, output, and internal signals from the DUV. This information is also used by the monitor to generate functional coverage data. The input signals provide a coverage estimate of the stimulus, whereas the internal and output signals provide an estimate of the coverage of functionality. In advanced verification environments, the stimulus components also use coverage data obtained from internal DUV probes or DUV inputs by the monitor to steer the stimulus generation. This is called coverage-directed stimulus generation. In SystemVerilog and Vera these monitors are called coverage groups.

Monitors also provide post-simulation information, which could be used for debugging purposes. A trade-off has to be made while probing input, output, and internal signals of the DUV for collecting information to be passed on to the checker or scoreboard by limiting the internal probes to ease excessive reliance on the design for information.

6.2.3   Checker

The purpose of checkers is to verify the correctness of the DUV. A checker is one of the more complicated components to implement, since it has to have an incisive notion of correctness of the design. There are three types of checker from an implementation perspective. The protocol and functional correctness checkers communicate with the input and output monitors or scoreboard for the information they need while the assertions hook up to the DUV directly for sampling input, output, and internal signals.

6.2.3.1   Protocol checker

A protocol checker is supposed to signal an error if the DUV does not follow the underlying protocol. These protocol compliance checkers are implemented based on the design specification and not based on the intent of the designer. They may obtain stimulus information from external monitors or the scoreboard or have embedded monitors for that purpose in order to predict functional results independently. They use this information to compute the expected reference value of outputs independent of the DUV execution, which are then compared with the collected outputs of the DUV. In the case of a miscompare, the checker flags an error and halts the simulation. Since there may be many requests and interacting stimuli in a single test case, a checker has to correlate input requests with output responses.

6.2.3.2   Functional correctness checker

Functional correctness checkers have to be aware of internal intricacies of the DUV to the extent that they are very often considered as reference models. Traditionally, functional correctness was checked by analyzing test case traces by hand and by observing results on the DUV outputs. This was possible since the earlier designs were simpler and had fewer corner conditions and complex interactions. But with increasing design complexity, automated checker components, which are like miniature reference models, are essential. They are very similar to protocol checkers in terms of communication with the rest of the testbench. In addition to access to the inputs and outputs of DUV they may also need internal signals for computing the expected results, depending on the complexity of the underlying reference model. A trade-off has to be made in terms of dependency on internal signals, which represent reliance on designer's intent. This reliance cannot be completely avoided in this case because of the involved nature of functional correctness checks.

6.2.3.3   Assertions

Assertions can be considered as monitors as well as checkers. [5] They do not interact with the rest of the testbench other than the DUV. They monitor input, output, and internal signals of the DUV and check the design for correctness. One major difference between assertions and the other two checkers is that the former executes concurrently with the design while the latter do not. The assertions continuously monitor correctness of the DUV while it is executing, whereas the other two checkers check the outputs only at the end of the execution. The assertions are modeled to reflect desired properties, which should be exhibited by DUV during its execution. Any deviation from these properties causes a violation of the associated assertion and an error is flagged at the run time. Assertions afford a checking mechanism very tightly coupled to the design. However, the trade-off is to keep the DUV properties precise. If they become complex, then the assertions run the risk of replicating the DUV implementation.

6.2.4   Scoreboard

A scoreboard is a temporary holding location for information that a checker may require. A checker uses a scoreboard in two ways. The main difference between the two methods stems from which component acts as a reference model to check correctness. In this first method, the checker component contains the reference model. The scoreboard examines the inputs for transactions to occur, captures relevant information, and stores the information for later use. When the checker observes a condition on the outputs of the DUV, it fetches data from the scoreboard by making a call to it. The scoreboard implementation depends on the DUV functionality. If the DUV follows a FIFO protocol, the scoreboard would also follow the same protocol. If the DUV follows a complex queuing algorithm, then a complex function, such as a search based on port number, will have to be performed in the scoreboard to obtain the correct data. The returned data are used by the reference model in the checker to compute expected results, which are then compared with the DUV output signals.

In the second method, the scoreboard contains the reference model, which performs the expected result computation based on the observed input stimulus. The checker observes the DUV output events, queries the scoreboard for the expected data, and then performs the comparison. Either of the work-division schemes between checker and scoreboard is fine, so long as the choice of reference model placement remains consistent.

Sometimes, if the implementation is simple, the scoreboard also subsumes the checker component.

6.2.5   Stimulus

The stimulus components manipulate inputs to the DUV. These components ensure that the DUV is simulated through enough design states to verify its functionality adequately. The stimulus components typically consist of transaction generators, drivers, and irritators. A stimulus component also maintains a record of its activity for post-simulation analysis needed for test-case debugging.

6.2.5.1   Transaction generator

The transaction generators model the behavior of real design entities of the system of which the DUV is a subsystem. To create stimuli to be applied to the DUV, the transaction generator merely needs to model the interface inputs to the DUV, instead of modeling the entire behavior of these design entities. This interface is called a transaction interface. Only having to model the interface to be presented to the DUV without getting into the internal-logic-level details of the external design entities drastically reduces the complexity of development of a transaction generator, as well as simplifying the verification as a whole. The generator can produce stimulus in the following three ways:

- *Random* The stimuli are completely randomly generated from their value space using a pseudo-random generator.
- *Constrained random* The stimuli are constrained by specifying the relationship between the inputs or bounds on the values of the inputs. Sophisticated constraint solvers are used for this purpose. This method is used when the completely random method yields stimuli inadmissible by the DUV.
- *Directed* This method is used to generate stimuli to reach certain design states in the simulation that are very difficult to attain using the random and constrained-random methods.

The above mechanisms ensure detection of design errors in corner cases.

6.2.5.2   Driver

A driver takes as input the stimulus generated by the transaction generator and applies it to the inputs of the DUV. This whole process is called a transaction. A driver accurately models the interface protocol with the DUV. It is responsible for application of the inputs to the DUV in an appropriate sequence. It is also used to simulate the reset behavior by driving reset values of inputs into the DUV.

6.2.5.3   Irritator

Sometimes during the verification of complex DUVs, it is very difficult to simulate certain design states or, as they are called, corner cases. In these cases, in the interest of thorough verification, some internal signals of the design have to be forced to certain values in order for the DUV to execute the corner states. The mechanisms that inject these values onto the internal signals are called irritators. They also follow one of the three methods for value generation, as described in Section 6.2.5.1.

6.3   Verification plan

The Verification plan is a very important part of the verification process. It should be the first step in the verification process. The idea is for verification engineers to acquaint themselves thoroughly with the functionality of the DUV. A baseline verification plan consists of the following:

- A list of the design features to be verified,
- The details of correctness checks required to ensure that the DUV is correct with respect to those features,
- The details of coverage monitors required to make sure that those features are exercised in simulation.

The Verification plan is thoroughly reviewed with the design engineers so that the intricacies of the design choices are brought to the attention of the verification engineers. This allows the verification engineers to produce an unambiguous and faultless verification environment.

6.4   Case study

We will take up a design as an exercise and show through a sequence of steps how a verification environment is set up and a DUV is verified. We will demonstrate this with both SystemVerilog and Vera. These two languages have striking similarities, to the extent that most of the verification infrastructure code is reusable among the two. Unless it is explicitly mentioned, each code example discussed applies to System-Verilog as well as Vera; the filename extension will need to be changed to .vr from .sv, except in the case of HDL code, in which case it will be .sv in both these languages. Both SystemVerilog and Vera code examples are shown separately in the instances where they differ.

6.4.1   DUV

The example in Fig. 6.2 shows a typical interrupt controller. The controller takes interrupt information, such as interrupt event, interrupt event mask, and interrupt mask, as inputs and generates interrupt as an output. It also clears interrupt lines based

```
module  intCtrl (
// Inputs
clock, reset, Event, EventMask, ClrMode, StatRegRd,
StatRegWr, StatWrData, Mask,
// Outputs
Status, dynMask, intr);

input clock, reset;
input [3:0] Event;
input [3:0] EventMask;
input [3:0] Mask;
input [3:0] ClrMode;
input [3:0] StatWrData;
input  StatRegWr;
input  StatRegRd;

output [3:0] Status;
output reg [3:0] dynMask;
output reg intr;

integer i;

always @ (posedge clock or negedge reset)
   begin
   if (!reset) Status <= 0;
   else
      begin
      for (i=0; i<4; i++)
         begin
         if (Event[i] && EventMask[i])
            Status[i] <= 1'b1;
         else
            if (ClrMode[i])
               if (StatRegRd)
                  Status[i] <= 1'b0;
            else
               if (StatRegWr && !StatWrData[i])
                  Status[i] <= 1'b0;
         end
      end
   end
always @ (posedge clock or negedge reset)
   begin
   if (!reset) dynMask <= 0;
   else
      if (StatRegRd)
         dynMask <= dynMask | Status;
      else
         if (StatRegWr)
            dynMask <= dynMask & StatWrData;
   end

always @ (posedge clock or negedge reset)
   begin
   if (!reset) intr <= 1'b0;
   else
      if (((Status & ~dynMask) & Mask) == {4{1'b0}})
         intr <= 1'b0;
      else
         intr <= 1'b1;
   end
endmodule
```

Figure 6.2   A four-line interrupt controller (filename: intCtrl.sv)

on a read or a write operation on the interrupt status register. The Verilog design description always has three process blocks.

The controller has four interrupt lines, which allows for four different interrupt events. For signals with four bits, each bit corresponds to an interrupt line. The following is a description of the input signals:

- *Event* Each of the four bits represents a unique incoming interrupt event. A set bit signifies the presence of the corresponding event and vice versa.
- *EventMask* Each of the four bits represents masking information for an interrupt event. A set bit means that the corresponding event should be taken into account when computing the interrupt status, and vice versa.
- *ClrMode* Each of the four bits depicts the mode of clearing of an interrupt status bit pertaining to an interrupt event. A set bit would mean that the status bit should be cleared on a read of the interrupt status register. Otherwise, the status bit would be cleared on a write to the interrupt status register (a zero written to the corresponding bit).
- *StatRegRd* This indicates a read of the interrupt status register.
- *StatRegWr* This indicates a write to the interrupt status register.
- *StatWrData* Each of the four bits represents the data to be written to the corresponding bit of the interrupt status register.
- *Mask* Each of the four bits contains the masking information for an interrupt status bit. A set bit signifies that the corresponding interrupt status line should be taken into account for computation of the final interrupt.

The following is a description of the output signals:

- *Status* Each of the four bits represents whether an interrupt event has been recorded or not. A set bit indicates the presence of an event and vice versa.
- *dynMask* Each of the four bits represents dynamic masking information for an interrupt status bit. This is used to mask off a status bit that has already resulted in the generation of an interrupt. A set bit indicates that the corresponding interrupt status bit should be ignored in the course of computation of the final interrupt. These bits are cleared by a write to the interrupt status register, which sets its corresponding bits to one or zero.
- *intr* This is the final interrupt signal generated. The *dynMask* and the *Mask* values are ANDed with the contents of the interrupt status register. The *intr* is generated if any bit in the interrupt status register remains set after this computation.

The first process computes an AND of the interrupt event and the interrupt event mask information and sets the interrupt status lines in the interrupt status register if the corresponding bits remain set after the computation. It also administers the interrupt clearing information as to whether an interrupt line should be cleared on a read from or a write to the interrupt status register.

The second process generates the dynamic masking information *dynMask*, to be used in conjunction with the interrupt status register, while generating the interrupt signal so that an already existing event in the interrupt status register does not end up regenerating the interrupt.

The third process generates the final interrupt if any interrupt status register bits stay set after applying the masks, *dynMask* and *Mask*.

6.4.2   Verification plan

Table 6.1 represents a verification plan for the interrupt controller described above. The following plan is not meant to be exhaustive: it is aimed at giving the reader an idea of how one should go about developing such a plan. Coming up with a more detailed plan can be taken as an exercise by the reader.

6.4.3   Testbench

Figure 6.3 depicts the top-level HDL testbench, which applies to both SystemVerilog- and Vera-based verification environments. Line numbers 14 through 21 generate the clock and the reset such that the design is kept under reset for ten clock cycles before the transactions are applied. Line numbers 22 through 34 instantiate the DUV. Line number 35 instantiates the top-level SystemVerilog or Vera testbench, which is bound to the port interface instantiated in line numbers 36 through 48.

Figures 6.4(a) and (b) depict the port interface definition in SystemVerilog and Vera respectively. The purpose of this interface is to provide a mapping between the ports of the DUV and those of the testbench. One exception is the port *tr_no* which is added just to keep track of the input transactions.

Figure 6.5 shows the top level of the SystemVerilog testbench, which will be identical for Vera. The only difference will be in the file name, which will be tb.vr in the latter case, as opposed to tb.sv in the former. In lines 12 through 16, it instantiates the transaction generator, scoreboard, input, output, and coverage monitors. In lines 17 through 20, input and output monitoring tasks are invoked in parallel so that the data could be posted to the scoreboard. In the end, the task gen_trans is invoked to generate transactions.

Figure 6.6 presents the base object definition containing the data members that will be generated, manipulated, or observed by the testbench. Lines 2 through 8 represent inputs to the DUV. The *rand* type declaration allows them to be randomized. Lines 9 through 11 represent the outputs that are observed in the testbench. Line 12 depicts a variable *tr_no*, which is an artifact of the testbench to keep track of transactions, so that expected values of outputs could be compared with the computed values.

Figure 6.7 shows the code for the transaction generator. Lines 2 and 3 instantiate the transaction object and the driver. The "new" function in lines 4 through 8 connects the driver class to the defined port interface. The figure depicts three tasks to demonstrate generation of transactions in random, constrained-random, and direct modes.

The task *gen_random_trans* in lines 9 through 20 generates 20 completely random transactions. It instantiates a transaction object in each of the 20 iterations, and creates random values for each of the *rand* type data fields of the transaction object using a native SystemVerilog function called *randomize()*. It then stores the iteration number as the transaction number in the data field *tr_no*. The transaction number is tracked, to

Table 6.1 Verification plan

No.	Property or feature	Correctness	Coverage
1.	Status of an interrupt event in conjunction with the event masking data must be reflected in the interrupt status register	i. If there is an interrupt event and if the corresponding interrupt mask is set, then the event must be reflected in the interrupt status register. ii. If there is an interrupt event and if the corresponding interrupt mask is not set, then the event must not show up in the interrupt status register.	i. All four interrupt event types should be generated. ii. All interrupt event mask bits should be exercised. iii. Interrupt event masking and unmasking should be exercised for all interrupt events. (This will require cross-product coverage of the above two.)
2.	Interrupt status register must be cleared based on the clearing mode selected	i. Interrupt status register must be cleared on its read if clear on read mode is selected. ii Interrupt status register must be cleared only on a write of zero to it if clear on write mode is selected.	i. Clear on read and write modes both must be exercised for all the four interrupt status bits. ii. Read from the status register must occur. iii. Write to the status register must occur. iv. 0 and 1 must both be written to every bit of the status register. v. Read mode must be selected in conjunction with a read operation. (A cross-product of i and ii above.) vi. Write mode must be selected in conjunction with write operation. (A cross-product of i, iii, and iv above.)
3.	An interrupt event must not be able to generate an interrupt more than once unless the same event type is re-asserted.	i. If there is an interrupt status register read, the dynamic masking register must update itself so as to mask the status bits corresponding to active interrupt events. ii. If there is a write of 0 to the interrupt status register, the dynamic masking register must update itself so as to unmask the status bits corresponding to active interrupt events.	i. A read of the interrupt status register must occur. ii. A write of the interrupt status register must occur. iii. Interrupt status register bits must show both active and inactive interrupt events. iv. A write to the interrupt status register must result in status bits being both 0 and 1. v. A cross-product of i and iii above. vi. A cross-product of ii and iv above.
4.	The interrupt should be generated if an interrupt status register bit stays high after masking computation has been performed on it.	i. Interrupt must be generated if any interrupt status bit stays high after masking. ii. Interrupt must not be generated if none of the interrupt status bits stays high after masking.	i. Both scenarios of interrupt occurring and not occurring must be exercised. ii. Interrupt should be generated because of all interrupt events at least once.

```
 1 `include "intCtrl.sv"

 2 module intCtrl_tb ();
 3 wire [3:0] Event;
 4 wire [3:0] EventMask;
 5 wire [3:0] Mask;
 6 wire [3:0] ClrMode;
 7 wire [3:0] StatWrData;
 8 wire  StatRegWr;
 9 wire  StatRegRd;
10 wire [3:0] Status;
11 wire [3:0] dynMask;
12 wire intr;
13 reg clock, reset;

14 initial clock = 0;
15 always #1 clock = ~clock;
16 initial
17    begin
18    reset = 1;
19    repeat(10) @ (posedge clock);
20    reset = 0;
21    end

22 intCtrl intCtrl_inst(
23 .clock          (clock),
24 .reset          (reset),
25 .Event          (EventMask),
26 .ClrMode        (ClrMode),
27 .StatRegRd      (StatRegRd),
28 .StatRegWr      (StatRegWr),
29 .StatWrData     (StatWrData),
30 .Mask           (Mask),
31 .Status         (Status),
32 .dynMask        (dynMask),
33 .intr           (intr)
34 );

35 intCtrl_top top (ports);

36 intCtrl_interface ports(
37 .clock          (clock),
38 .reset          (reset),
39 .Event          (EventMask),
40 .ClrMode        (ClrMode),
41 .StatRegRd      (StatRegRd),
42 .StatRegWr      (StatRegWr),
43 .StatWrData     (StatWrData),
44 .Mask           (Mask),
45 .Status         (Status),
46 .dynMask        (dynMask),
47 .intr           (intr)
48 );

49 endmodule
```

Figure 6.3   HDL testbench top (filename: top.sv)

compare expected outputs with the obtained outputs in the scoreboard. It finally invokes the task *drive_intCtrl* from the driver class to propagate the transaction created to the DUV ports.

The task *gen_constr_trans* in lines 21 through 32 is quite similar to *gen_random_trans*, except for line number 27, where it adds a constraint such that none of the

```
 1 interface intCtrl_interface (
 3 input wire clock,
 4 input wire reset,
 5 input logic [3:0] Event,
 6 input logic [3:0] EventMask,
 7 input logic [3:0] Mask,
 8 input logic [3:0] ClrMode,
 9 input logic [3:0] StatWrData,
10 input logic  StatRegWr,
11 input logic  StatRegRd,
12 output logic [3:0] Status,
13 output logic [3:0] dynMask,
14 output logic intr,
15 input logic tr_no
16 );
17 endinterface
```

Figure 6.4(a)   Port interface for SystemVerilog (filename: intCtrl_interface.sv)

```
 1 interface intCtrl_interface (
 3 input  clock              CLOCK;
 4 input  reset              PHOLD #1;
 5 input  [3:0] Event        PHOLD #1;
 6 input  [3:0] EventMask    PHOLD #1;
 7 input  [3:0] Mask         PHOLD #1;
 8 input  [3:0] ClrMode      PHOLD #1;
 9 input  [3:0] StatWrData   PHOLD #1;
10 input  StatRegWr          PHOLD #1;
11 input  StatRegRd          PHOLD #1;
12 output [3:0] Status       PHOLD #1;
13 output [3:0] dynMask      PHOLD #1;
14 output intr               PHOLD #1;
15 input  tr_no              PHOLD #1;
16 );
17 endinterface
```

Figure 6.4(b)   Port interface for Vera (Filename: intCtrl_interface.vr)

20 randomly created constraints have all the four interrupt lines masked. This is just a sample constraint; more constraints can be added to tune the pattern generation in a desired way.

The task *gen_direct_trans* in lines 33 through 46 generates an extremely directed single transaction aimed at verifying a very specific scenario. It exercises the scenario where:

• There is an interrupt event in line 2 (second bit from left in *Event*),
• None of the events or the interrupts is masked,
• The interrupt status register is set to clear on a write to it,
• A write of 0 to the corresponding bit of status register is issued.

Finally, the generated transaction is driven to the DUV ports by the driver task *drive_intCtrl*.

```
1 `include "intCtrl_interface.sv"

2 program intCtrl_top (intCtrl_interface ports);
3 `include "intCtrl_base_object.sv"
4 `include "intCtrl_driver.sv"
5 `include "intCtrl_xgen.sv"
6 `include "intCtrl_scoreboard.sv"
7 `include "intCtrl_ip_monitor.sv"
8 `include "intCtrl_op_monitor.sv"
9 `include "intCtrl_cov.sv

10 initial
11   begin
12   intCtrl_scoreboard sb    = new ();
13   intCtrl_ip_monitor ipm  = new (sb, ports);
14   intCtrl_op_monitor opm  = new (sb, ports);
15   intCtrl_xgen xgen   = new (ports);
16   intCtrl_cov cov  = new (ports);
17   fork
18     ipm.input_monitor();
19     opm.output_monitor();
20   join_none
21   xgen.gen_random_trans();
22   repeat (20) @ (posedge ports.clock);
23   end
24 endprogram
```

Figure 6.5 SystemVerilog testbench top (filename: tb.sv)

```
1  class intCtrl_base_object;
2  rand bit [3:0] Event;
3  rand bit [3:0] EventMask;
4  rand bit [3:0] Mask;
5  rand bit [3:0] ClrMode;
6  rand bit [3:0] StatWrData;
7  rand bit  StatRegWr;
8  rand bit  StatRegRd;
9  bit [3:0] Status;
10 bit [3:0] dynMask;
11 bit [3:0] intr;
12 bit tr_no;
13 endclass
```

Figure 6.6 Base object (filename: object.sv)

The transactions can be generated in random, constrained-random, or directed modes, depending on the choice of task used in the top-level SystemVerilog testbench in line number 21 of Figure 6.7.

Figure 6.8 shows the driver that delivers the generated stimulus to the DUV ports. In lines 3 through 14, the port interface is initialized as would be the case when starting from reset. The task *drive_intCtrl* in lines 15 through 27 connects the data fields of the transaction object to corresponding ports of the DUV through the port interface.

```
1 class intCtrl_xgen;
2 intCtrl_base_object intCtrl_object;
3 intCtrl_driver intCtrl_driver;

4 function new (virtual intCtrl_ports ports);
5    begin
6    inCtrl_driver = new (ports);
7    end
8 endfunction

9 task gen_random_trans();
10   begin
11   int i, result;
12   for (i = 0; i < 20; i++)
13      begin
14      intCtrl_object = new ();
15      result = intCtrl_object.randomize();
16      intCtrl_object.tr_no = i;
17      intCtrl_driver.drive_intCtrl(intCtrl_object);_
18      end
19   end
20 endtask
21 task gen_constr_trans();
22   begin
23   int i, result;
24   for (i = 0; i < 20; i++)
25      begin
26      intCtrl_object = new ();
27      result = intCtrl_object.randomize() with
                {Mask[3:0] != 4'b0000};
28      intCtrl_object.tr_no = i;
29      intCtrl_driver.drive_intCtrl(intCtrl_object);_
30      end
31   end
32 endtask
33 task gen_direct_trans();
34   begin
35   intCtrl_object = new ();
36   intCtrl_object.Event = 4'b0010;
37   intCtrl_object.Event Mask = 4'b1111;
38   intCtrl_object.Mask = 4'b1111;
39   intCtrl_object.ClrMode = 1'b0;
40   intCtrl_object.StatWrData = 4'b0000;
41   intCtrl_object.StatRegWr = 1'b1;
42   intCtrl_object.StatRegRd = 1'b0;
43   intCtrl_object.tr_no = 0;
44   intCtrl_driver.drive_intCtrl(intCtrl_object);_
45   end
46 endtask

48 endclass
```

Figure 6.7   Transaction generator (filename: intCtrl_xgen.sv)

Figure 6.9 shows the input monitor, which collects transaction data from input ports of the DUV. Lines 5 through 10 show instantiations of the scoreboard and port interface with the new function. The task *input_monitor* in lines 11 through 29 executes in a non-terminating while loop, creates a transaction object, and stores the inputs in the corresponding fields of the object. It also copies the transaction number in the *tr_no* field, so that the transaction could be tracked against the corresponding outputs when posted on the scoreboard using the scoreboard task *post_input*.

```
 1 class intCtrl_driver;
 2 virtual intCtrl_ports ports;

 3 function new (virtual intCtrl_ports ports);
 4   begin
 5   this.ports = ports;
 6   ports.Event = 0;
 7   ports.EventMask = 0;
 8   ports.Mask = 0;
 9   ports.ClrMode = 0;
10   ports.StatWrData = 0;
11   ports.StatRegWr = 0;
12   ports.StatRegRd = 0;
13   end
14 endfunction

15 task drive_intCtrl (intCtrl_base_object object);
16   begin
17   @ (posedge ports.clock);
18   ports.Event = object.Event;
19   ports.Event Mask = object.EventMask;
20   ports.Mask = object.Mask;
21   ports.ClrMode = object.ClrMode;
22   ports.StatWrData = object.StatWrData;
23   ports.StatRegWr = object.StatRegWr;
24   ports.StatRegRd = object.StatRegRd;
25   $display("Interrupt stimulus received");
26   end
27 endtask
28 endclass
```

Figure 6.8   Driver (filename: intCtrl_driver.sv)

Figure 6.10 shows the output monitor, which is used to collect transaction data from output ports of the DUV. Lines 5 through 10 show instantiations of the scoreboard and port interface with the *new* function. The task *output_monitor* in lines 11 through 25 executes in a non-terminating while loop, creates a transaction object, and stores the outputs in the corresponding fields of the object. It also copies the transaction number in the *tr_no* field so that the transaction could be tracked against the corresponding inputs when posted on the scoreboard using the scoreboard task *post_output*.

Figure 6.11 presents the assertions corresponding to the correctness check items in the third column of Table 6.1. For example, the assertion corresponding to the correctness item i. under feature number 1. is named *property_1_i*. The assertions are shown in lines 14 through 27. The assertion file is included in the top-level HDL testbench and the assertion module is bound to the DUV with the bind statement in line number 29.

Figure 6.12 shows the coverage monitors implemented in SystemVerilog. Each of the cover points or cross products in lines 8 through 32 represents a coverage item in the fourth column of Table 6.1. For example, the cover point corresponding to the coverage item i. under feature number 1. is named *cov_1_i*. The cover points sample the signals from the port interface. The cross-product coverage items are implemented as cross products between the coverage points on the individual signals involved, as shown in line number 11.

```
 1 class intCtrl_ip_monitor;
 2 intCtrl_base_object intCtrl_object;
 3 intCtrl_scoreboard sb;
 4 virtual intCtrl_ports ports;

 5 function new (intCtrl_scoreboard, virtual intCtrl_ports ports);
 6   begin
 7   this.sb = sb;
 8   this.ports = ports;
 9   end
10 endfunction

11 task input_monitor ();
12   begin
13   while (1)
14     begin
15     @ (posedge ports.clock);
16     intCtrl_object = new ();
17     $display("Input monitor : Transaction %d Stored", ports.tr_no);
18     intCtrl_object.Event = ports.Event;
19     intCtrl_object.EventMask = ports.EventMask;
20     intCtrl_object.Mask = ports.Mask;
21     intCtrl_object.ClrMode = ports.ClrMode;
22     intCtrl_object.StatWrData = ports.StatWrData;
23     intCtrl_object.StatRegWr = ports.StatRegWr;
24     intCtrl_object.StatRegRd = ports.StatRegRd;
25     intCtrl_object.tr_no = ports.tr_no;
26     sb.post_input(intCtrl_object);
27     end
28   end
29 endtask
30 endclass
```

Figure 6.9   Input monitor (filename: intCtrl_ip_monitor.sv)

```
 1 class intCtrl_op_monitor;
 2 intCtrl_base_object intCtrl_object;
 3 intCtrl_scoreboard sb;
 4 virtual intCtrl_ports ports;

 5 function new (intCtrl_scoreboard, virtual intCtrl_ports ports);
 6   begin
 7   this.sb = sb;
 8   this.ports = ports;
 9   end
10 endfunction

11 task output_monitor ();
12   begin
13   while (1)
14     begin
15     @ (negedge ports.clock);
16     intCtrl_object = new ();
17     $display("Output monitor : Transaction %d Retrieved", ports.tr_no);
18     intCtrl_object.tr_no = ports.tr_no;
19     intCtrl_object.Status = ports.Status;
20     intCtrl_object.dynMask = ports.dynMask;
21     intCtrl_object.intr = ports.intr;
22     sb.post_output(intCtrl_object);
23     end
24   end
25 endtask
26 endclass
```

Figure 6.10   Output monitor (filename: intCtrl_op_monitor.sv)

```
 1 module  intCtrl_assert (
 2 input clock, reset,
 3 input [3:0] Event,
 4 input [3:0] EventMask,
 5 input [3:0] Mask,
 6 input [3:0] ClrMode,
 7 input [3:0] StatWrData,
 8 input  StatRegWr,
 9 input  StatRegRd,
10 input [3:0] Status,
11 input [3:0] dynMask,
12 input intr
13 );

14 generate
15   genvar bit_count;
16     for (bit_count=0; bit_count<4; bit_count++)
17       begin
18         property_1_i: assert property (@(posedge clock) disable iff      (!reset)
(Event[bit_count] && EventMask[bit_count]) |=> Status[bit_count];
19         property_1_ii: assert property (@(posedge clock) disable iff (!reset)
!(Event[bit_count] && EventMask[bit_count]) |=> !Status[bit_count];
20         property_2_i: assert property (@(posedge clock) disable iff (!reset)
(ClrMode[bit_count] && StatRegRd) |=> !Status[bit_count];
21         property_2_ii: assert property (@(posedge clock) disable iff (!reset)
(!ClrMode[bit_count] && StatRegWr && !StatWrData[bit_count] ) |=> !Status[bit_count];
22         property_3_i: assert property (@(posedge clock) disable iff (!reset) (StatRegRd
&& Status[bit_count]) |=> dynMask[bit_count];
23         property_3_ii: assert property (@(posedge clock) disable iff (!reset) (StatRegWr
&& !StatWrData[bit_count]) |=> !dynMask[bit_count];
24         property_4_i: assert property (@(posedge clock) disable iff (!reset) (|((Status &
~dynMask) & Mask) |=> intr;
25         property_4_ii: assert property (@(posedge clock) disable iff (!reset) (!|((Status &
~dynMask) & Mask)  |=> !intr;
26       end
27 endgenerate
28 endmodule
29 bind intCtrl intCtrl_assert intCtrl_assert_inst (.*);
```

Figure 6.11   Correctness check with assertions

For cases where the cross product of individual samples becomes prohibitively large, only interesting cases are monitored. For example, in lines 17 through 20, a cross product of signals ClrMode, StatRegWr, and StatWrData is being taken. However, only the cross product items where StatRegWr is 1 are monitored.

6.5   Summary

This chapter demonstrates the use of SystemVerilog and Vera as HDLs for functional verification. We do not specifically promote the use of these particular languages, although they both have excellent features for verification. We present the use of these

```
 1 class intCtrl_cov;
 2 virtual intCtrl_ports ports;
 3 function new (virtual intCtrl_ports ports);
 4   begin
 5   this.ports = ports;
 6   end
 7 endfunction
 8 covergroup cvg @ (posedge clock);
 9    cov_1_i: coverpoint ports.Event;
10    cov_1_ii: coverpoint ports.EventMask;
11    cov_1_iii: cross cov_1_i, cov_1_ii;
12    cov_2_i: coverpoint ports.ClrMode;
13    cov_2_ii: coverpoint ports.StatRegRd;
14    cov_2_iii: coverpoint ports.StatRegWr;
15    cov_2_iv: coverpoint ports.StatWrData;
16    cov_2_v: cross cov_2_i, cov_2_ii;
17    cov_2_vi: cross cov_2_i, cov_2_iii, cov_2_iv
18      {
19          bins b1 = binsof(cov_2_iii) intersect {1};
20      }
21    cov_3_iii: coverpoint ports.Status;
22    cov_3_v: cross cov_2_ii, cov_3_iii
23      {
24          bins b2 = binsof(cov_2_ii) intersect {1};
25      }
26    cov_4_i: coverpoint ports.intr;
27    cov_4_ii: cross cov_4_i, cov_1_i
28      {
29          bins b2 = binsof(cov_4_i) intersect {1} &&
30                     ! binsof(cov_1_i) intersect {0};
31      }
32 endgroup
33 endclass
```

Figure 6.12   Coverage groups in SystemVerilog (filename: intCtrl_cov.sv)

languages here to assist in the application of some of the verification ideas presented in this book with real simulation tools. This chapter gives only an introductory view of the use of SystemVerilog and Vera.

6.6   References

[1] *IEEE Standard for SystemVerilog: Unified Hardware Design, Specification and Verification Language*, IEEE Std. 1800–2005.

[2] F. Haque, J. Michelson, and K. Khan (2001). *The Art of Verification with VERA*. Verification Central.

[3] I. G. Harris (2005). Hardware/software covalidation. *IEE Proceedings on Computers and Digital Techniques*, **152(3)**:380–392.

[4] B. Wile, J. C. Goss, and W. Roesner (2005). *Comprehensive Functional Verification*. Morgan Kaufman.

[5] H. Foster, A. Krolnik, and D. Lacey (2004). *Assertion-Based Design*. 2nd edn. Kluwer.

7 Decision diagrams for verification

Maciej Ciesielski, Dhiraj K. Pradhan, and Abusaleh M. Jabir

7.1   Introduction

Having matured over the years, formal design verification methods, such as theorem proving, property and model checking, and equivalence checking, have found increasing application in industry. Canonical graph-based representations, such as binary decision diagrams (BDDs), [1] binary moment diagrams (BMDs), [2] and their variants, play an important role in the development of software tools for verification. While these techniques are quite mature at the structural level, the high-level verification models are only now being developed. The main difficulty is that such verification must span several levels of design abstraction. Verification of arithmetic designs is particularly difficult because of the disparity in the representations on the different design levels and the complexity of logic involved.

This chapter addresses verification based on canonical data structures. It presents several canonical, graph-based representations that are used in formal verification, and, in particular, in equivalence checking of combinational designs specified at different levels of abstraction. These representations are commonly known as *decision diagrams*, even though not all of them are actually decision-based forms. They are graph-based structures whose nodes represent the variables and whose directed edges represent the result of the decomposition of the function with respect to the individual variables. Particular attention is given to arithmetic and word-level representations.

An important common feature of all these representations is canonicity, which is essential in combinational equivalence checking. A form is canonical if the representation of a function in that form is unique. Canonical graph-based representations make it possible to check whether two combinational functions are equivalent by checking whether their graph-based representations are isomorphic. Isomorphism can be checked in constant time, once the representation has been constructed, by testing if the two functions share the same root of the diagram.

The canonical diagrams can be fully characterized by the following basic properties, described in detail in this chapter:

1. *Decomposition principle*, which defines the types of function that can be modeled by the diagram and the underlying decomposition method. They include binary

Practical Design Verification, eds. Dhiraj K. Pradhan and Ian G. Harris. Published by Cambridge University Press. © Cambridge University Press 2009.

decomposition for Boolean functions, some form of multi-valued decomposition for integer-valued functions, and moment decomposition or other non-binary expansions for arithmetic functions.

2. *Simplification rules* that make the diagram minimal and irredundant, hence canonical. Different rules apply to different types of diagrams.

3. *Composition algorithms*, which, given graph-based representations for functions F and G, specify how to construct a similar representation for function $F < op > G$, where $< op >$ represents an operation defined for the given application domain (Boolean, arithmetic, finite field, etc.). The composition algorithms, commonly known as APPLY algorithms, recursively apply the given operation $< op >$ to the decomposed functions, depending on the type of functions and operations allowed.

One of the most well known and commonly used canonical diagram representations is binary decision diagram (BDD). [1] Binary decision diagrams are based on the well-known Shannon (or more accurately, Boole) function expansion, which decomposes the function into two co-factors, $f(x = 0)$ and $f(x = 1)$. Each sub-graph resulting from such a decomposition can be viewed as a decision ($x = 0$ or $x = 1$) taken at a decomposing variable, justifying the name *decision diagram*. Binary decision diagrams have been developed for Boolean functions and logic circuits represented at the bit level and used extensively in representing and verifying bit-level designs, such as control and random logic. Thanks to their compact, canonical form they truly revolutionized the field of combinational verification and logic synthesis and found applications in many other fields, such as satisfiability, testing, and synthesis. However, because of their exponential worst-case size complexity, they have had limited success in modeling and verifying RTL designs with significant arithmetic components, especially with multipliers.

Another canonical form described in this chapter is a binary moment diagram (BMD), [2] developed specifically for arithmetic functions. Binary moment diagrams are based on a moment decomposition principle, which treats an arithmetic function as a linear function with Boolean inputs and integer (or real-valued) outputs. The two sub-functions resulting from the decomposition represent the two moments (constant and linear) of the function, rather than a "decision." For this reason, BMDs do not technically belong to a category of decision diagrams but form a class of their own. Binary moment diagrams find important applications in verifying arithmetic designs with bit-level inputs and integer outputs.

Two newer types of diagrams, called Taylor expansion diagrams (TEDs) and finite-field decision diagrams (FFDDs), have recently been introduced to address the need for a more abstract design representation, with inputs and outputs allowed to take either integer or discrete (finite-field) values. Both of these diagrams can be thought of as extensions of BDDs and BMDs, with inputs and outputs represented as symbolic variables. The two diagrams differ in arithmetic representation of the data (infinite-precision integer vs. finite-field arithmetic) and the type of decomposition used (Taylor expansion vs. multi-valued Galois field (GF) decomposition).

Taylor expansion diagrams (TEDs) [3] are based on Taylor expansions of polynomial representation of the computation expressed in the design. Both inputs and outputs are treated as infinite-precision integers (or real numbers) and are represented by symbolic variables. The power of abstraction, combined with canonicity and compactness, makes the TED particularly attractive for verification of designs specified at the behavioral and algorithmic levels, such as datapaths and signal processing systems. Computations performed by those designs can often be expressed as polynomials and can be efficiently represented with TEDs, with memory requirements several orders of magnitude smaller than those of other known representations. Taylor expansion diagrams can also serve as a vehicle to transform the initial functional representation of the design into a structural representation in the form of a dataflow graph (DFG); as such, they are applicable to behavioral synthesis, or, more specifically, to behavioral transformations, which can also be used in verification.

Finite-field decision diagrams (FFDDs) [4] are an extension of multiple-terminal decision diagrams, but with inputs and outputs represented in the finite-field (also called Galois field, GF) arithmetic rather than in the integer domain. Finite-field representation has numerous applications in cryptography, error-control systems, fault-tolerant designs, and digital signal processing. Finite-field decision diagrams allow the simulation and verification of such systems to be performed more efficiently on a higher level of abstraction. The verification can be performed either at the bit or at the word level; it is not restricted to word boundaries and can be used to model and verify any combination of output bits.

7.2   Decision diagrams

Binary decision diagrams (BDD) have emerged as the representation of choice for many applications, ranging from representation of Boolean function, through verification and satisfiability, to logic synthesis. Even though BDDs (albeit under a different name) have been known since the late 1950s, it was the seminal work of Bryant [1] that brought to light their importance as canonical representations for Boolean logic. This section briefly reviews the basic theory and algorithms of BDDs, taken from multiple sources. [1,5–7]

7.2.1   Binary decision diagrams (BDDs)

A binary decision diagram is a graph-based data structure, which represents a set of binary-valued decisions, culminating at an overall decision that can be either true or false. Specifically, a BDD is a directed acyclic graph (DAG) whose nodes represent the decisions, and edges represent the decision types (true or false). The final decision evaluated at the root represents the overall function encoded by the BDD. Ordered and reduced BDDs are irredundant and canonical, i.e., a representation of a function in that form is unique. Formally, a BDD is defined as follows:

DEFINITION 7.1 *A binary decision diagram (BDD) is a rooted directed acyclic graph G(V,E) with a set of nodes V and a set of edges E. The vertex set V contains two types of vertex:*

- *Two terminal nodes (leaves), corresponding to constants 0 and 1.*
- *A set of variable nodes {u}, each associated with a Boolean variable $v = var(u)$. Each node has exactly two outgoing edges, pointing to two child functions, low(u) and high(u). (In the figures, the two child edges are represented as dotted and solid lines, respectively.)*

The function of node $u \in V$, associated with variable $v = var(u)$, is given by $f^u = \bar{v} \cdot low(u) + v \cdot high(u)$, where low(u) and high(u) are the functions of the low and high children of u, respectively. In particular, the function evaluated at the root represents the logic function encoded in the BDD.

The decomposition principle

The above definition basically states that a BDD is based on a Shannon (Boole) expansion of function f, applied recursively to its variables. That is,

$$f^u = \bar{v} \cdot f_{\bar{v}} + v \cdot f_v, \tag{7.1}$$

where $f_{\bar{v}} = low(u)$ and $f_v = high(u)$ are the negative and positive co-factors of f with respect to the decomposing variable v.

DEFINITION 7.2 *A BDD is ordered (denoted OBDD) if on all paths from the root to its terminal nodes, the variables appear in the same linear order: $x_1 < x_2 < \ldots > x_n$. Furthermore, the OBDD is reduced (denoted ROBDD) if it satisfies two properties:*

1. *(Irredundancy) No variable node u has identical low and high children, i.e., low $(u) \neq high(u)$.*
2. *(Uniqueness) No two distinct nodes u and v have the same variable name and the same low and high children. That is, $var(u) = var(v)$, $low(u) = low(v)$, $high(u) = high(v) \Rightarrow u = v$.*

The above definition provides the *reduction rules* for BDDs: rule 1 removes redundant nodes with identical low and high children; rule 2 merges isomorphic sub-graphs. The resulting ROBDDs form an irredundant representation, i.e., no two nodes of the ROBDD represent the same Boolean function. Two ROBDDs are *isomorphic* if there is a one-to-one mapping between the vertex sets that preserves adjacency, indices, and leaf values. Thus, two isomorphic ROBDDs represent the same function. Conversely, two Boolean expressions that represent the same logic function have isomorphic ROBDDs for a given ordering of variables. In this sense, ROBDDs form a canonical representation.

The following lemma, from Bryant, states the canonicity of ROBDDs. [1]

LEMMA 7.3 *For any Boolean function f there is exactly one ROBDD with root node u and variable order $x_1 < x_2 < \ldots < x_n$ such that $f^u = f(x_1, x_2, \ldots, x_n)$.*

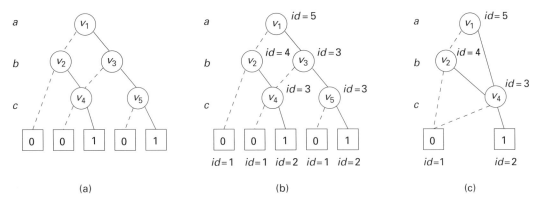

Figure 7.1   Construction of an ROBDD for $f = (a + b)c$: (a) OBDD for the variable order a,b,c; (b) OBDD with unique identifiers; (c) ROBDD for variable order a,b,c.

BDD construction

An algorithm has been proposed by Bryant to reduce OBDD. The resulting diagram, ROBDD, is irredundant, minimal, and canonical. The algorithm visits the OBDD bottom up, from the leaf nodes to the root, and labels each vertex $v \in V$ with an identifier $id(v)$. The reduction rules are then applied to remove redundant nodes and merge isomorphic sub-graphs. As a result, an ROBDD is identified by a subset of vertices with different identifiers.

The algorithm is illustrated in Fig. 7.1, taken from [5] for function $f = (a + b)c$. An OBDD is constructed from the original expression, as shown in Fig 7.1(a). Then, the nodes of the OBDD are labeled with identifiers, as a function of the variable name and their children. First, the leaf nodes (0 and 1) are labeled with identifiers $id = 1$ and $id = 2$, respectively. Then the vertices v_4, v_5 on the bottom-most level, corresponding to variable c, are labeled with their identifiers. In this case both nodes are assigned the same identifier, $id = 3$, since they correspond to the same variable and have children with the same identifiers. They are replaced by a single node, v_4 (visited first), added to the ROBDD. Next, the algorithm visits vertices v_2, v_3, associated with variable b. Vertex v_2 is assigned identifier $id = 4$ and is added to the ROBDD. The left (low) and right (high) children of node v_3 have the same identifier, so v_3 inherits their identifier and is discarded as redundant. Finally, the root v_1 associated with variable a is visited and assigned the identifier $id = 5$. It is added to the ROBDD as a unique node with this identifier. The resulting ROBDD is shown in Fig. 7.1(c).

In practice, ROBDDs are built directly from a Boolean formula, avoiding the reduction step and possible memory overflow problems. This approach is based on applying the Shannon decomposition, $f = \bar{v} \cdot f_{\bar{v}} + v \cdot f_v$, iteratively to the variables of the formula in a predetermined order. Canonicity and minimality of such constructed ROBDDs are accomplished by using a hash table, called the *unique table*, which contains a key for each vertex of an ROBDD, and which uniquely identifies the function associated with that node. The key is a triple, composed of the variable name and the identifiers of the low and high children. The unique table is constructed from the bottom up. When a new node is considered for the addition to the ROBDD, a lookup in the table determines whether another vertex in the table already implements the same functionality by comparing the keys. If this is the case, the pointer of the new node is set to the one existing in the table; otherwise a new entry is made in the table for the new node. In this way, no redundant nodes are added to the table and the table represents an ROBDD. The run-time complexity of this and other ROBDD construction algorithms is $O(2^n)$, where n is the number of variables. Similarly, the size of the ROBDD is, in the worst case, exponential. The details of the construction of an ROBDD can be found in [1,5].

Binary decision diagrams provide a compact representation of Boolean logic. Each path of the BDD from root to terminal node 1 represents a product term (on-set cube) of the function encoded in the BDD. It is computed as a product of variables, along the path, at their respective polarity. For example, for a BDD in Fig. 7.1, a path $\{v_1, v_2, v_4, 1\}$ corresponds to the product term $\bar{a}bc$. The logic function encoded in a BDD is then evaluated as a logical sum (OR) of product terms associated with the on-paths. Similarly, a path from the root to the terminal node 0 represents a complement of the function. This feature is useful for function complementation, which can be achieved in constant time by simply exchanging the 0 and 1 terminal nodes.

Reduced ordered binary decision diagrams can naturally represent multiple-output functions by modeling them as ROBDDs with shared sub-graphs. In the following, we will refer to ROBDD simply as BDD, since some ordering of the variables is always imposed on the BDD, and the ROBDD must be reduced in order to be canonical.

BDD composition – the APPLY algorithm

Another way of constructing a BDD for a given Boolean expression is to compose BDDs of its sub-expressions using Boolean connectives, such as AND, OR, or XOR. The algorithm that performs such a composition is known as the APPLY algorithm.

The basic idea comes, again, from the recursive application of Shannon expansion theorem for arbitrary binary operator $<op>$:

$$f<op>g = \bar{v}(f_{\bar{v}}<op>g_{\bar{v}}) + v(f_v<op>g_v). \qquad (7.2)$$

Starting with the topmost variable v in the two functions, the formula is applied recursively to all the variables in the order that they appear in their respective BDDs (f and g must have compatible ordering for the algorithm to work). If v is the top variable of f and g, then the operator $< op >$ is applied to their respective co-factors. If f does not depend on f_v, then $f_v = f_{\bar{v}} = f$ and the co-factor is the function itself. The worst-case complexity of the APPLY algorithms is $O(n_1 \cdot n_2)$, where n_1 and n_2 are the number of variables in the two BDDs.

Using the above algorithm, one can construct a BDD for an arbitrary Boolean network or a gate-level netlist. First, a trivial BDD is built for the variables representing primary inputs, and then BDDs of each expression or logic gate are constructed from the BDDs of their immediate inputs, in topological order, from primary inputs to primary outputs.

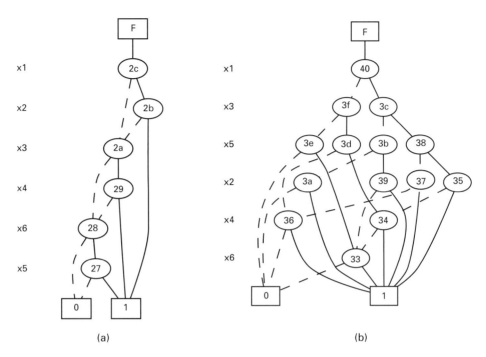

Figure 7.2   Effect of variable ordering on BDD size for function $F = x_1 x_2 + x_3 x_4 + x_5 x_6$: (a) for variable order $x_1, x_2, x_3, x_4, x_5, x_6$, (b) for variable order $x_1, x_3, x_5, x_2, x_4, x_6$.

Operations on BDDs can be done in polynomial time of their size (number of nodes). However, the real complexity is hidden in their construction, which is expensive in both time and space. Binary decision diagrams can be exponential in size and cause memory explosion, especially for designs containing arithmetic functions, such as multipliers (BDDs cannot be built for multipliers larger than about 16×16 bits).

Variable ordering

The size of a BDD strongly depends on the ordering of the variables. The size of the BDD (measured as the number of nodes) is, in the worst case, exponential in the number of variables. Reduced ordered binary decision diagrams representing adder functions are particularly sensitive to the variable order; they can have exponential size in the worst case and a linear size in the best case. There are functions (such as multipliers), whose BDD size is exponential regardless of the ordering. Furthermore, there are functions for which the sum of products (SOP) or product of sums (POS) forms are more compact than the BDDs. Many constraint functions of covering problems fall into this category. Figure 7.2 shows two BDDs for function $F = x_1 x_2 + x_3 x_4 + x_5 x_6$ constructed with two different orderings of variables, lexicographical, and interleaved. One can see a significant difference in BDD size.

While the variable ordering problem is NP-complete, efficient heuristic variable ordering algorithms exist, based on both static (related to lexicographical) and dynamic ordering (swapping variables on two adjacent levels). [8]

Extensions

Several extensions have been proposed for BDDs. One of them makes use of *complemented edges* by labeling BDD edges with complement attributes. This feature makes it possible to represent a function and its complement as a single sub-graph with two edges coming into the root of the sub-graph, one with positive polarity and the other with negative polarity. In general, BDDs with complemented edges result in a smaller BDD size and provide a means to complement a BDD in constant time. To maintain the canonicity, certain restrictions are imposed on the placement of the complemented edges. Namely, only *low* edges, corresponding to negative co-factors, may be assigned complement attributes. Notice that for BDDs with complemented edges, only one constant function (1) and, hence, only one terminal node (leaf 1) is needed, since 0 can be derived from its complement.

Applications and limitations of BDDs

Owing to their compactness, canonicity, and ease of manipulation, BDDs have found numerous applications in design, synthesis, verification, and testing of digital designs. In general, BDDs are an efficient data structure for storing and evaluating Boolean functions and discrete structures. Large sets of discrete elements can be encoded in binary and compactly represented as characteristic functions in BDDs. Binary decision diagrams are particularly handy in representing transition relations of product machines for the purpose of sequential equivalence checking using state traversal. [7]

In particular, BDDs have found widespread application in a number of verification problems, including combinational equivalence checking, [9] implicit state enumeration and FSM traversal, [7,10] symbolic model checking, [11,12] and test vector generation. Their biggest claim to fame comes from their applications to *combinational equivalence checking*. Once two logic functions are represented by their respective ROBDDs (with the same variable order), one can test whether the two functions are equivalent by testing whether their ROBDDs are isomorphic. In practice, checking for equivalence between two functions is performed by constructing a single, multi-rooted BDD, rather than checking for graph isomorphism. The two functions are equivalent if they share the same root. This test can be made in constant time, once the BDD is built for the two functions.

Logic equivalence can be illustrated with the BDD shown in Figure 7.1(c). The ROBDD in the figure, constructed for function $f = (a + b)c$, also represents function $g = a \cdot c + b \cdot c$, as well as a number of other equivalent functions, all having the same BDD for a fixed variable order. As mentioned earlier, BDDs can be built for an arbitrary gate-level network or a multiple-output Boolean function. Such created BDDs can then be used to check the equivalence of the netlist to another netlist, or to the initial Boolean specification of the design.

Another obvious application of BDDs is *satisfiability* (SAT). A Boolean function is satisfiable if there exists an assignment of Boolean values to its variables that makes the function true ($f = 1$). Many verification, synthesis, and optimization problems can be reduced to the SAT problem. Being decision diagrams, BDDs can be used to solve the SAT problem in linear time in its size. Once the formula to be satisfied is converted to a BDD, the BDD is traversed to find one or more paths from the root to the terminal node 1. A satisfying solution exists, so long as the BDD is not empty. This important feature of decision diagrams finds its application in deterministic *test generation*, used in simulation-based verification. A target assignment, not adequately covered by semi-random or directed simulation, is specified and solved using BDD-based SAT.

The BDD-based approach to the SAT problem can be illustrated with the example in Figure 7.1(c). Two satisfying solutions for $f = 1$, corresponding to the paths from the root to node 1, are $\{ac\}$ and $\{\bar{a}bc\}$.

A special case of SAT is related to finding a satisfying assignment for $f = 0$. A notion of *easily invertible* form was introduced by Bryant to denote a representation for which it is always possible to find a zero of the function (solve for $f = 0$) in polynomial time. [13] Clearly, BDDs are easily convertible functions, since one can find a solution to the problem by tracing the path from the root to the terminal node 0. Another special case of SAT involves testing for *tautology*, i.e., testing if the function is identical to 1 for all assignments of Boolean variables. This can be done in constant time by testing if BDD for the function is reduced to constant 1.

Several efficient implementations of software programs supporting BDDs have been developed for a wide set of purposes. [14] One of the most popular packages, available on the World Wide Web, is the CUDD package. [15]

In summary, BDDs have been very successful in verifying control-dominated applications and are a part of a number of formal verification systems, such as SMV [12] and VIS. [16] However, as the designs have grown in size and complexity, the size-explosion problems of BDDs have limited their scope. Furthermore, their use in designs containing large arithmetic data-path units have been limited due to prohibitive memory requirements, especially for large multipliers.

7.2.2   Beyond BDDs

In an attempt to obtain a more compact representation for Boolean functions, different flavors of Boolean decomposition have been tried. These diagrams, collectively known as *decision diagrams*, are still based on a "point-wise" binary decomposition, but use a different interpretation of the diagram nodes.

One such representation is based on the XOR (exclusive OR) decomposition:

$$f = f_{\bar{x}} \oplus x f_{\Delta x} = f_x \oplus \bar{x} f_{\Delta x}, \qquad (7.3)$$

also known as Red–Miller or Davio decomposition. Here, $f_{\Delta x}$ denotes the Boolean difference of function f w.r.t. variable x, i.e., $f_{\Delta x} = f_x \oplus f_{\bar{x}}$, where $\oplus$ represents an XOR operation.

Ordered functional decision diagrams (*OFDDs*) [17] are based on such a decomposition. This representation is analogous to that of OBDDs, except that the two outgoing arcs at each node represent the negative co-factor and the Boolean difference of the function w.r.t. the node variable. As with OBDDs, OFDD representation is canonical and many operations can be implemented with algorithms of polynomial complexity. However, several important features differentiate the two representations.

First, different reduction rules are applied to make the graph canonical. Second, the evaluations of a function on an OFDD involves more than tracing a path. In particular, for a node variable x, both sub-graphs must be evaluated and an XOR computed. Such an evaluation can be performed in linear time in the number of nodes by a post-order traversal of the graph. An interesting feature of OFDDs is that, for certain classes of function (in particular, arithmetic functions based on XORs), OFDDs are exponentially more compact than ROBDDs, but the reverse is also true. To obtain the advantages of each representation, Drechsler *et al.* proposed a hybrid form, called ordered Kronecker FDD (OKFDD). [18] In this representation, each variable can use any of the three decompositions given by Eqs. 7.1–7.3, potentially leading to a reasonable reduction in the graph size.

Another variant of BDD representation, called zero-suppressed BDDs (ZBDDs), was developed by Minato for solving combinatorial problems. [19] Zero-suppressed BDDs are particularly suitable for applications involving sparse sets of bit vectors. It can be shown that ZBDDs reduce the size of the representation of a set of n-bit vectors over OBDDs by at most a factor of n. In practice, the reduction is large enough to have a significant impact.

Numerous attempts have been made to extend the capabilities of BDDs to target arithmetic circuits and designs with word-level specifications. This requires extending the concept of Boolean function representation to integer and real-valued functions over Boolean variables. The resulting graph-based representations for functions with a Boolean domain and an integer range are commonly known as *word-level decision diagrams* (WLDDs). [20,21]

One straightforward way to represent numeric-valued functions is to use the branching structure of a BDD, but to allow arbitrary values on the terminal nodes. Such a representation is referred to as a multi-terminal BDD (MTBDD) [22] or algebraic decision diagram (ADD). [23] Evaluating an MTBDD or ADD for a given variable assignment is similar to evaluating a BDD. However, MTBDDs are inefficient in representing functions yielding values over a large range, as this requires a large number of terminal nodes and results in a large number of paths (MTBDDs tend to be trees rather than graphs).

For such applications, alternative representations have been proposed, such as edge-valued BDDs (EVBDDs). [24] These forms incorporate numeric weights on the edges, to allow greater sharing of sub-graphs and to reduce the size of the overall representation. Evaluating a function represented by an EVBDD involves tracing the path determined by the variable assignment and adding the products of variable values along the path, weighted by the corresponding edge weights. This representation grows linearly as the number of bits, a major improvement over MTBDDs. However, the overhead for storing and normalizing the edge weights to make the representation canonical makes them less efficient. There are classes of function, such as arithmetic functions, for which EVBDD has unacceptable size complexity. In particular, the EVBDD representation for integer multipliers, $F = X \cdot Y$, grows exponentially with the number of bits of its operands. A good review of WLDDs can be found in [20,21].

In the next section, another type of word-level diagram is described, based on a different, non-pointwise decomposition principle.

7.3   Binary moment diagrams (BMDs)

An alternative approach to representing numeric functions, especially those encountered in arithmetic circuits, involves changing the function decomposition with respect to its variables.

The decomposition principle

Binary moment diagrams (BMDs), introduced by Bryant [13], use a modified Shannon's expansion, in which a Boolean variable is treated as a binary $(0,1)$ integer variable. The complement of x is modeled as $\bar{x} = 1 - x$, and the terms of the expansion are regrouped around variable x, resulting in the following formula:

$$\begin{aligned} f(x) &= (1 - x) \cdot f_{\bar{x}} + x.f_x \\ &= f_{\bar{x}} + x \cdot (f_x - f_{\bar{x}}) \\ &= f_{\bar{x}} + x \cdot f_{\Delta x}, \end{aligned} \tag{7.4}$$

where $\cdot$, $+$, and $-$ denote multiplication, addition, and subtraction, respectively. The above decomposition is termed *moment* decomposition; $f_{\bar{x}}$ is the *constant moment*, and $f_{\Delta x} = f_x - f_{\bar{x}}$ is the *linear moment*. In this form, f can be viewed as a *linear function* in x, with f_x as the constant term, and $f_{\Delta x}$ as the linear coefficient of f (the partial derivative of f with respect to x). This expansion still relies on the assumption that variable x is Boolean, i.e., evaluates to either 0 or 1. However, it departs from a point-wise, decision-based decomposition and performs the decomposition of a linear function based on its first two moments.

Each node of a BMD describes a function in terms of its moment decomposition with respect to the variable labeling the node, as shown in Fig. 7.3(a). The two outgoing arcs from each node denote the constant moment (shown as dashed lines) and the first moment (solid lines) of the function w.r.t. the decomposing variable. Part (b) of the figure shows the BMD representation of the unsigned integer $X = 4x_2 + 2x_1 + x_0$ encoded with $n = 3$ bits. The constants in the terminal nodes of the BMD can be moved to their edges, and represented as edge-weights, as shown in Fig. 7.3(c). The resulting

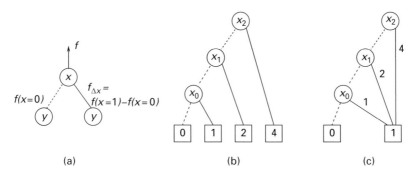

(a)   (b)   (c)

Figure 7.3   Binary moment diagrams: (a) the moment decomposition principle; (b) BMD for binary encoded integer $X = 4x_2 + 2x_1 + x_0$; (c) *BMD for X.

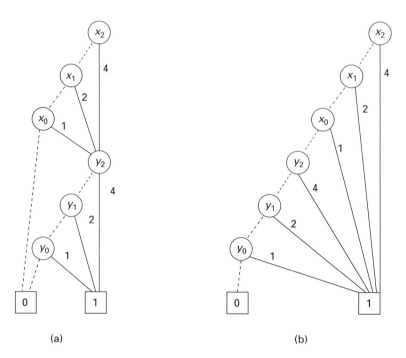

Figure 7.4   *BMD representations for word levels operations: (a) sum $X + Y$; (b) product $X\, Y$.

diagram is termed a multiplicative binary moment diagram, or *BMD. The term *multiplicative* derives from the fact that, when evaluating a function along a path from root to one of its terminal nodes, the weights combine multiplicatively along the path.

Similarly to BDDs, a function encoded in a *BMD is evaluated by adding the terms encoded in the paths. However, two major features differentiate *BMDs from the decision diagrams discussed earlier:

1. *BMDs are *not* decision diagrams, since they are based on moment decomposition, rather than a point-wise Shannon expansion.
2. *BMDs are *multiplicative* diagrams, in the sense that each path from a root to a terminal node is a product of the variables labeling the nodes and the edge weights along the path.

Figure 7.4 shows *BMDs for addition and multiplication expressed at word levels. Note that the size of *BMDs for these operations grows linearly with the word size n.

Reduction rules
Each node in the *BMD is represented as a triple $< v,\ low(v),\ high(v) >$, with two weights associated with the constant and linear moments, $w_0(v),\ w_1(v)$. It is assumed that the set of variables is totally ordered, as in a BDD. To maintain the canonical form, certain reduction rules must be imposed on the *BMD during node creation and weight manipulation (normalization). In principle, these rules are similar to those in

BDDs, but must follow the rules of regular algebra $(+,\cdot)$ rather than Boolean algebra $(\vee,\wedge)$.

1. *Irredundancy* When a linear moment at node v is 0, the function at the node evaluates to its constant moment, i.e., does not depend on v. In this case node v is redundant and is removed. (Note that this rule differs from the redundancy reduction rule for BDD).
2. *Uniqueness* This rule is similar to that of BDD: any two nodes indexed by the same variable and having the same two moments represent the same function and are merged in the BMD into a single node. This rule, however, is applied after the normalization, described next.

Normalization

Several rules for manipulating edge weights are imposed on the graph to make the graph canonical. For non-zero values of the linear moment at node v, the weights of its two edges are normalized by factoring out the greatest common divisor (gcd) of the argument weights $w = gcd(w_0(v),w_1(v))$, which is then pushed to the root edge of node v. By convention, the sign of the extracted weight must match the sign of the constant moment; in this way, gcd always returns a non-negative value. Normalization is performed from the bottom up, from the leaf nodes to the root. Each normalized node is stored in the hash table, where each entry is indexed by a key composed of the variable and the two moments. Duplicate entries are automatically removed, resulting in an irredundant, minimal, and canonical representation.

As with BDDs, the *BMD representation of a function depends on the variable order, but *BMDs are much less sensitive to variable ordering than BDDs.

The APPLY algorithms

As with BDDs, *BMDs are constructed by starting with base functions, corresponding to constants and single variables, and then building more complex functions according to some operation. Algorithms similar to the APPLY algorithm for BDDs have been proposed. However, while there is a single APPLY algorithm for BDDs for an arbitrary Boolean operator, *BMDs require algorithms tailored specifically for the individual operations, such as ADD, SUB and MULT. [13] In general,

$$f<op>g = (f<op>g)_{\bar{x}} + x(f<op>g)\Delta x, \tag{7.5}$$

where

$$(f<op>b)_{\bar{x}} = (f_{\bar{x}}<op>g_{\bar{x}}), \tag{7.6}$$

and

$$\begin{aligned}
(f<op>g)_{\Delta x} &= (f<op>g)_x - (f<op>g)_{\bar{x}} \\
&= (f_x<op>g_x) - (f_{\bar{x}} - g_{\bar{x}}) \\
&= ((f_x + f_{\Delta x}) <op> (g_x + g_{\Delta x}) - (f_{\bar{x}} - g_{\bar{x}})).
\end{aligned} \tag{7.7}$$

However, in the case of the multiply operation, special attention must be paid because of the introduction of the term containing x^2.

$$f \cdot g = (f_{\bar{x}} + x \cdot f_{\Delta x}) \cdot (g_{\bar{x}} + x \cdot g_{\Delta x})$$
$$= f_{\bar{x}} \cdot g_{\bar{x}} + x \cdot (f_{\bar{x}} \cdot g_{\Delta x} + f_{\Delta x} \cdot g_{\bar{x}}) + x^2 \cdot f_{\Delta x} \cdot g_{\Delta x}. \tag{7.8}$$

The multiply operation must be linearized by replacing x^2 with x, since x is a Boolean variable. This gives the following result for multiplication:

$$f \cdot g = f_{\bar{x}} \cdot g_{\bar{x}} + x \cdot (f_{\bar{x}} \cdot g_{\Delta x} + f_{\Delta x} \cdot g_{\bar{x}} + f_{\Delta x} \cdot g_{\Delta x}). \tag{7.9}$$

The APPLY algorithms proceed by traversing the argument graphs and recursively apply the operation to the sub-graphs. To reduce the number of recursive calls, a hash table is maintained, keyed by the arguments of the previous calls, allowing the program to reuse previous computations.

Unlike operations on BDDs, which have run-time complexities that are polynomial in the number of variables, most operations on *BMDs potentially have exponential complexity. However, as demonstrated by Bryant, these exponential cases do not arise in practical applications. [13] Furthermore, the size of the arguments is significantly smaller in word-level applications than in bit-level applications, resulting in reasonable run-times.

For word-level expressions $(X+Y)$ and $(X*Y)$, where X and Y are n-bit vectors, the *BMD representation is linear in the number of bits n. Also, function c^X, where c is a constant, has a linear size representation in *BMD. However, the size of *BMD for X^k is $O(n^k)$. Thus, for high-degree polynomials defined over words with large bit widths, as commonly encountered in many DSP applications, filters, etc., *BMD remains an expensive representation.

Boolean logic

A *BMD can be adapted to also represent Boolean logic, which is important for designs with Boolean connectives. The following equations are used to model Boolean logic:

$$\text{NOT}: \quad \bar{x} = (1 - x), \tag{7.10}$$

$$\text{AND}: \quad x \wedge y = \quad x \cdot y, \tag{7.11}$$

$$\text{OR}: \quad x \vee y = \quad x + y - x \cdot y, \tag{7.12}$$

$$\text{XOR}: \quad x \oplus y = \quad x + y - 2x \cdot y. \tag{7.13}$$

Figure 7.5 shows *BMD representations for these basic Boolean operators. [2] In the diagrams, x and y are Boolean variables represented by binary variables, and $+$ and $\cdot$ represent algebraic operators of ADD and MULT, respectively. The resulting functions are 0,1 integer functions.

Multiplicative binary moment diagrams provide a concise representation of functions defined over bit vectors, or words of data, having a numeric representation. In particular they can efficiently encode integer-valued functions defined over binary-encoded words,

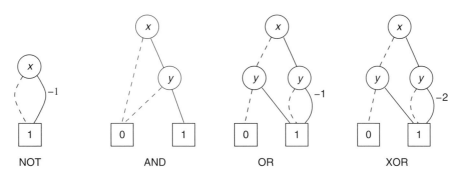

NOT AND OR XOR

Figure 7.5 *BMD representation for Boolean operators: (a) NOT: $\bar{x} = (1 - x)$; (b) AND: $x \wedge y = xy$; (c) OR: $x \vee y = x + y - xy$; (d) XOR: $x \oplus y = x + y - 2x$

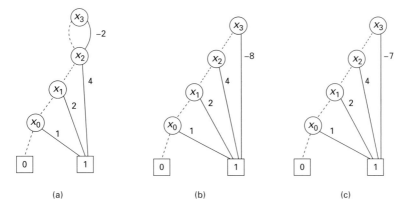

(a) (b) (c)

Figure 7.6 *BMD representations for word-level operations: (a) sign magnitude; (b) twos complement; (c) ones complement.

$X = \sum_i 2^i x_i$, where each $x_i = 0$ or 1. Figure 7.6 shows examples of *BMD representation for signed integers using several sign schemes (signed magnitude, ones complement, and twos complement). All commonly used encodings can be similarly represented.

Applications to word-level verification

Multiplicative binary moment diagrams have been successfully used in formal verification of arithmetic circuits. Figure 7.7 illustrates an approach to arithmetic circuit verification proposed in [13,25]. The goal is to prove a correspondence between a logic circuit, represented by a vector of Boolean functions f, and the design specification, represented by a word-level function f. The inputs to the Boolean circuit f are vectors of Boolean signals, $x_1, x_2, \ldots, x_k$; the inputs to the specification function F are word-level signals (symbolic variables) $X_1, X_2, \ldots, X_k$. To compare the two designs, each of the Boolean vectors x_i is transformed into a word-level signal X_i using an encoding function $Enc_i(x_i)$, and connected to the appropriate input of F. An encoding function simply provides the interpretation of the bit vectors. An example of such an encoding function (in this case, unsigned integer) is shown in Figure 7.3. Similarly, the output of

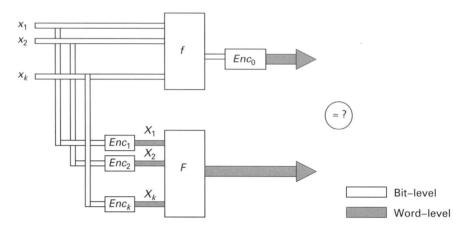

Figure 7.7   General verification problem: prove correspondence between a word-level specification and bit-level implementation

the logic circuit f is transformed to a word-level function using an encoding function Enc_0. The general task of verification is then to prove the equivalence between the circuit output, interpreted as a word, and the output of the word-level specification.

$$Enc_0(f(x_1, \ldots, x_k)) = F(Enc_1(x_1), \ldots, Enc_k(x_k)). \tag{7.14}$$

Multiplicative binary moment diagrams can provide a suitable data structure for this form of verification. Taylor expansion diagrams, described in the next section, can be used for the final comparison at the word-level.

A serious limitation of *BMDs is that they cannot be used for solving SAT problems. This is because *BMDs are multiplicative diagrams, i.e., the weights combine multiplicatively along the path from the terminal node to the root. Solving the integer-valued SAT problem in this structure is equivalent to solving the integer factorization problem. They are also not "easily invertible," as defined earlier in the context of the decision diagrams.

Several variants of *BMD representation have been proposed in the literature. Chen and Bryant introduced multiplicative power hybrid decision diagrams (PHDDs) that allow floating point arithmetic to be handled. [26] This is the only known form that supports floating point operation in a graph without introducing rational numbers, by representing the mantissa and exponent as connected sub-graphs. However, the size of the graph, even for the adder function, grows exponentially with the size of the exponent size.

Drechsler *et al.* extended *BMDs to a form called K*BMD to make the decomposition more efficient in terms of the graph size. This is done by admitting multiple decomposition types and allowing both additive and multiplicative edge weights. [27] However, a set of restrictions imposed on the edge weights to make it canonical makes such a graph difficult to construct. The K*BMD is characterized by linear complexity of the word-level operations for sum, product, and c^X, and can represent X^k in $O(n^{k-1})$ nodes. As we will see in the next section, this result can be further improved with TEDs, [3] which offer linear size complexity for this and other word-level operations.

7.4   Taylor expansion diagrams (TEDs)

Before formally introducing TEDs, we briefly review previous work and recent advances in word-level equivalence checking and the supporting symbolic representations.

7.4.1   Related work

In the realm of high-level design verification, the issue of abstraction of symbolic, word-level computations has received a lot of attention. This is visible in theorem-proving techniques, automated decision procedures for Presburger arithmetic, [28,29] techniques using algebraic manipulation, [30] symbolic simulation, [31] or in the decision procedures that use a combination of theories. [32,33] *Term rewriting* systems, particularly those used for hardware verification [34–36] also represent computations in high-level symbolic forms. These representations and verification techniques, however, do not rely on canonical forms. For example, verification techniques using term rewriting are based on rewrite rules that lead to normal forms. Such forms may produce false negatives, which may be difficult to analyze and resolve.

Various forms of *high-level logic* have been used to represent and verify high-level design specifications. Such representations are mostly based on quantifier-free fragments of first-order logic. The research that deserves particular mention includes: the logic of equality with uninterpreted functions (EUF) [37] and with memories (PEUFM), [38,39] and the logic of counter arithmetic with lambda expressions and uninterpreted functions (CLU). [40] These logics are often transformed into canonical representations, such as BDDs and BMDs, or into SAT instances or other normal forms. [41,42] To avoid exponential explosion of BDDs, equivalence verification is generally performed by transforming high-level logic description of the design into propositional logic formulas [33,38,40] and employing satisfiability tools [43,44] for testing the validity of the formulas. While these techniques have been successful in the verification of control logic and pipelined microprocessors, they have found limited application in the verification of large datapath designs.

Word-level ATPG techniques [45–49] have also been used for RTL and behavioral verification. However, their applications are generally geared toward simulation, functional vector generation or assertion property checking, but not so much toward high-level equivalence verification of arithmetic designs.

Symbolic algebra methods

Many computations encountered in behavioral design specifications can be represented in terms of polynomials. This includes digital signal and image processing designs, digital filter designs, and designs that employ complex transformations, such as DCT, DFT, WHT, etc. Polynomial representations of discrete functions have been explored in the literature long before the advent of contemporary canonical graph-based representations. In particular, Taylor's expansion of Boolean functions has been studied in [50, 51]. However, these works mostly targeted classical switching theory problems: logic minimization, functional decomposition, fault detection, etc. The issue of abstraction of bit-vectors and symbolic representation of computations for high-level synthesis and formal verification was not their focus.

Commercial symbolic algebra tools, such as Maple, [52] Mathematica, [53] and MatLab, [54] use advanced symbolic algebra methods to perform efficient manipulation of mathematical expressions, including fast multiplication, factorization, etc. However, despite the unquestionable effectiveness of these methods for classical mathematical applications, they are less effective in modeling large scale digital circuits and systems, and, in particular, in polynomial verification. For example, symbolic algebra tools offered by Mathematica and the like cannot unequivocally determine the *equivalence* of two polynomials. The equivalence is checked by subjecting each polynomial to a series of *expand* operations and comparing the coefficients of the two polynomials ordered lexicographically. As stated in the manual of Mathematica 5, *"There is no general way to find out whether an arbitrary pair of mathematical expressions are equal."*[53] Furthermore, Mathematica, *"Cannot guarantee that any finite sequence of transformations will take any two arbitrarily chosen expressions to a standard form."*

In contrast, the TED data structure described here provides an important support for equivalence verification by offering a canonical representation for multi-variate polynomials.

Equivalence checking

Equivalence checking has been researched thoroughly and there is a vast amount of literature on the topic, including satisfiability (SAT) approaches, [45,48,49,55,56] verification of arithmetic on bit-level, [57–59] symbolic approaches, and others. [60–66]

A typical approach to equivalence checking (EC), employed by industrial tools, involves identifying structural equivalences or *similarities* between pairs of points (called *cut points*) in the two designs. The portions of designs identified as having equivalent cut points are removed from the design and the EC verification is repeated on the reduced designs. However, the main difficulty lies in identifying such cut points in designs described in different levels (e.g., RTL and algorithmic). Another challenge in EC verification comes from structural optimizations, employed by behavioral or high-level synthesis (such as factorization, resource sharing, change of order of operators, operator merging, etc.), which reduce the level of similarity between the candidate cut points. The next section provides a motivating example for the development of symbolic equivalence techniques based on functional, rather than structural, approach and the associated canonical representation.

7.4.2   Motivation

The following example, shown in Fig. 7.8(a) and (b), taken from the Synopsys technical bulletin, [67] illustrates the perceived difficulty of functional verification of arithmetic designs in the case of combinational transformation, called resource sharing. Resource sharing transforms the netlist by moving the operators to maximize sharing of the resources, in this case the multiplication. The arithmetic proof engine

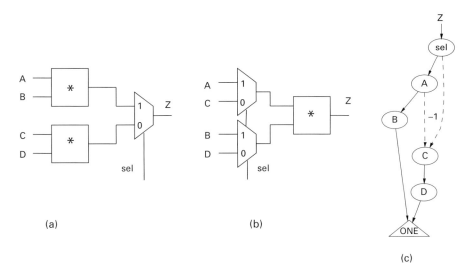

Figure 7.8 Verification of resource sharing: (a) $z = sel? (A{\cdot}B) : (C{\cdot}D)$; (b) $z = (sel? A : C) \cdot (sel ? B : D)$; (c) canonical TED showing functional equivalence of the two structures: $z = A \cdot B \cdot sel + C \cdot D \cdot (1 - sel) = (A \cdot sel + C \cdot (1 - sel)) \cdot (B \cdot sel + D \cdot (1 - sel))$ for $sel = 0,1$

(APF) of Synopsys' Formality tool cannot solve this problem using cut points because internally equivalent points are lost during such a transformation.

This problem can be solved, however, by generating symbolic expressions for the original and the transformed forms in a canonical form, and proving that they are equivalent. Namely, the function computed by the original design shown in Fig. 7.8(a) can be written as

$$z = A \cdot B \cdot sel + C \cdot D \cdot (1 - sel), \tag{7.15}$$

while the design in Fig. 7.8(b) can be expressed as

$$(A \cdot sel + C \cdot (1 - sel)) \cdot (B \cdot sel + D \cdot (1 - sel)). \tag{7.16}$$

Since sel is a binary variable, $sel^2 = sel$ and $sel \cdot (1 - sel) = 0$, and the above expression reduces to Eq. 7.16. These expressions can be captured by a canonical data structure with symbolic input variables A, B, C, D, sel. Such a diagram is shown in Fig. 7.8(c). The equivalence of the two designs can be verified by testing whether the diagrams corresponding to the two designs are isomorphic, which is the case here (only one graph is shown). This is the main idea behind Taylor expansion diagrams, described next. Note that, unlike BMDs, this diagram represents the designs with arbitrary bit width; that is the designs can be verified for equivalence regardless of their word sizes, assuming infinite-precision arithmetic.

7.4.3 The Taylor series expansion

A known limitation of all decision and moment diagram representations is that *word-level* computations, such as $A + B$, require the function to be decomposed with respect

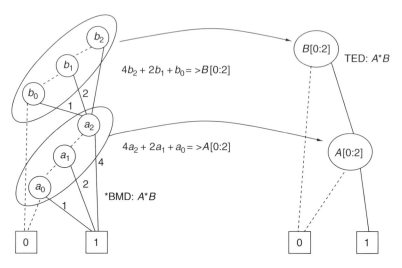

Figure 7.9   Abstraction of bit-level variables into algebraic symbols for $F = A \cdot B$

to *bit-level* variables $A[k]$, $B[k]$. Such an expansion creates a large number of variables in the respective diagram framework and requires excessive memory and time to operate upon them. To represent and process the HDL description of a large design efficiently, it is desirable to treat the word-level variables as *algebraic symbols*, expanding them into their bit-level components only when necessary.

Consider the *BMD for $A \cdot B$, shown in Fig. 7.9, which depicts the decomposition with respect to the bits of A and B. It would be desirable to group the nodes corresponding to the individual bits of these variables to *abstract* the integer variables they represent, and use the abstracted variables directly in the design. Figure 7.9 depicts the idea of such a *symbolic abstraction* of variables from their bit-level components.

To achieve the type of abstracted representation depicted above, one can rewrite the moment decomposition $f = f_{\bar{x}} + x \cdot (f_x - f_{\bar{x}})$ as $f = f(x = 0) + x \cdot \frac{\partial (f)}{\partial x}$. This equation resembles a truncated *Taylor series expansion* of the linear function f with respect to x. By allowing x to take integer values, the binary moment decomposition can be generalized to a Taylor series expansion, where integer variables do not need to be expanded into bits.

In this approach, an algebraic, multi-variate expression, $f(x,y,\ldots)$, can be viewed as a continuous, differentiable function over a real domain. It can be decomposed using the Taylor series expansion with respect to variable x as follows [68]:

$$f(x) = \sum_{k=0}^{\infty} \frac{1}{k!}(x - x_0)^k f^k(x_0) = f(x_0) + x f'(x_0) + \frac{1}{2}x^2 f''(x_0) + \ldots, \quad (7.17)$$

where $f'(x_0)$, $f''(x_0)$, etc., are first-, second-, and higher-order derivatives of f with respect to x, evaluated at $x_0 = 0$. The derivatives of f evaluated at $x = 0$ are independent of variable x, and can be further decomposed w.r.t. the remaining variables, one variable at a time. The resulting recursive decomposition can be represented by a decomposition diagram, called the *Taylor expansion diagram*, or TED.

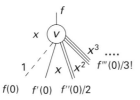

Figure 7.10   A decomposition node in a TED

The Taylor series expansion can be used to represent computations over integer and Boolean variables, commonly encountered in HDL descriptions. Arithmetic functions and dataflow portions of those designs can be expressed as multi-variate polynomials of finite degree, for which the Taylor series is finite.

DEFINITION 7.4 *The Taylor expansion diagram*, or *TED*, *is a directed acyclic graph* (Φ, V, E, T), *representing a multi-variate polynomial expression* Φ, *where* V *is the set of nodes,* E *is the set of directed edges, and* T *is the set of terminal nodes in the graph. Every node* $v \in V$ *has an index* var(v) *which identifies the decomposing variable. The function at node* v *is determined by the Taylor series expansion at* x = var(v) = 0, *according to Equation 7.17. The number of edges emanating from node* v *is equal to the number of non-empty derivatives of* f *(including* f(0)*) w.r.t. variable* var(v). *Each edge points to a sub-graph whose function evaluates to the respective derivative of the function with respect to* var(v). *Each sub-graph is recursively defined as TED w.r.t. the remaining variables. Terminal nodes evaluate as constants.*

Starting from the *root*, the decomposition is applied recursively to the subsequent children nodes. The internal nodes are in one-to-one correspondence with the successive derivatives of function f w.r.t. variable x evaluated at $x = 0$. Figure 7.10 depicts one-level decomposition of function f at variable x. The kth derivative of a function rooted at node v with $var(v) = x$ is referred to as a *k-child* of v; $f(x = 0)$ is a 0-child, $f'(x = 0)$ is a 1-child, $\frac{1}{2!}f''(x = 0)$ is a 2-child, etc. We shall also refer to the corresponding arcs as *0-edge* (dotted), *1-edge* (solid), *2-edge* (double), etc.

Example 7.1 Figure 7.11 shows the construction of a TED for the algebraic expression $F = A^2 + AB + 2AC + 2BC$. Let the ordering of variables be A, B, C. The decomposition is performed first with respect to variable A. The constant term of the Taylor expansion is $F(A = 0) = 2 \cdot B \cdot C$. The linear term of the expansion gives $F'(A = 0) = B + 2C$; the quadratic term is $\frac{1}{2} \cdot F''(A = 0) = \frac{1}{2} \cdot 2 = 1$. This decomposition is depicted in Fig. 7.11 (a). Now the Taylor series expansion is applied recursively to the resulting terms with respect to variable B, as shown in Fig. 7.11(b), and subsequently with respect to variable C. The resulting diagram is depicted in Fig. 7.11(c), and its final reduced and normalized version (to be explained in Section 7.4.4) is shown in Fig. 7.11(d). The function encoded by the TED can be evaluated by adding all the paths, from the non-zero terminal nodes to the root, each path being a product of the variables in their respective powers and the edge weights, resulting in $F = A^2 + AB + 2AC + 2BC$.

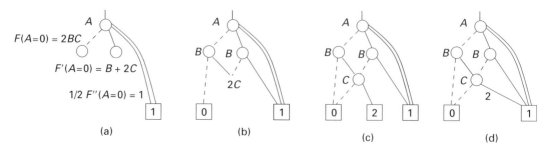

Figure 7.11 Construction of a TED for $F = A^2 + AB + 2AC + 2BC$: (a)–(c) decomposition w.r.t. individual variables; (d) normalized TED

Using the terminology of computer algebra, [69] TED employs a *sparse recursive representation*, where a multivariate polynomial $p(x_1, \ldots, x_n)$ is represented as:

$$p(x_1, \ldots, x_n) = \sum_{i=0}^{m} p_i(x_1, \ldots, x_{n-1}) x_n^i. \qquad (7.18)$$

The individual polynomials $p_i(x_1, \ldots, x_{n-1})$ can be viewed as *coefficients* of the leading variable x_n at the decomposition level corresponding to x_n. By construction, the sparse form stores only non-zero polynomials as the nodes of the TED.

7.4.4   Reduction and normalization

It is possible to reduce the size of an ordered TED further by a process of TED *reduction* and *normalization*. Analogous to BDDs and *BMDs, Taylor expansion diagrams can be reduced by removing redundant nodes and merging isomorphic sub-graphs. In general, a node is redundant if it can be removed from the graph, and its incoming edges can be redirected to the nodes pointed to by the outgoing edges of the node, without changing the function represented by the diagram.

DEFINITION 7.5 *A TED node is* redundant *if all of its non-0 edges are connected to terminal 0.*

If node v contains only a constant term (0-edge), the function computed at that node does not depend on the variable $var(v)$, associated with the node. Moreover, if all the edges at node v point to the terminal node 0, the function computed at the node evaluates to zero. In both cases, the parent of node v is reconnected to the 0-child of v, as depicted in Fig. 7.12.

The identification and merging of isomorphic sub-graphs in a TED are analogous to that of BDDs and *BMDs. Two TEDs are considered *isomorphic* if they match in both their structure and their attributes; i.e. if there is a one-to-one mapping between the vertex sets and the edge sets of the two graphs that preserve vertex adjacency, edge labels, and terminal leaf values. By construction, two isomorphic TEDs represent the same function. To make the TED canonical, any redundancy in the graph must be eliminated and the graph must be reduced. The reduction process entails merging the isomorphic sub-graphs and removing redundant nodes.

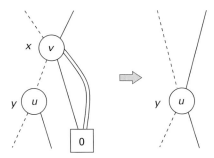

Figure 7.12   Removal of redundant node with only a constant term edge

DEFINITION 7.6 *A Taylor expansion diagram is* reduced *if it contains no redundant nodes and has no distinct vertices v and v', such that the sub-graphs rooted at v and v' are isomorphic. In other words, each node of the reduced TED must be unique.*

It is possible to reduce the graph further by exploiting the sharing of common sub-expressions by performing *normalization*, similar to the method described for *BMDs. [2] The normalization procedure starts by moving the numeric values from the non-zero terminal nodes to the terminal edges, where they are assigned as edge *weights*. This is shown in Fig. 7.11(d) and Fig. 7.13(b). By doing this, the terminal node holds constant 1. This operation applies to all terminal edges with terminal nodes holding values different from 1 or 0. As a result, only terminal nodes 1 and 0 are needed in the graph. The weights at the terminal edges may be further propagated to the upper edges of the graph, depending on their relative values. The TED normalization process that accomplishes this is defined as follows.

DEFINITION 7.7 *A reduced, ordered TED representation is normalized when:*

- *The weights assigned to the edges spanning out of a given node are relatively prime,*
- *Numeric value 0 appears only in the terminal nodes,*
- *The graph contains no more than two terminal nodes, one each for 0 and 1.*

By ensuring that the weights assigned to the edges spanning out of a node are relatively prime, the extraction of common sub-graphs is enabled. Enforcing the rule that none of the edges is allowed zero weight is required for the canonization of the diagram. When all the edge weights have been propagated up to the edges, only the values 0 and 1 can reside in the terminal nodes.

The normalization of the TED representation is illustrated by an example in Fig. 7.13. First, as shown in Fig. 7.13(b), the constants (6, 5) are moved from terminal nodes to terminal edges. These weights are then propagated up along the linear edges to the edges rooted at nodes associated with variable B, see Fig. 7.13(c). At this point the isomorphic sub-graphs $(B + C)$ are identified at the nodes of B and the graph is subsequently reduced by merging the isomorphic sub-graphs, as shown in Fig. 7.13(d).

It can be shown that the normalization operation can reduce the size of a TED exponentially. Conversely, transforming a normalized TED to a non-normalized TED can, in the worst case, result in an exponential increase in the graph size. This result follows directly from the concepts of normalization of BDDs and BMDs. [2]

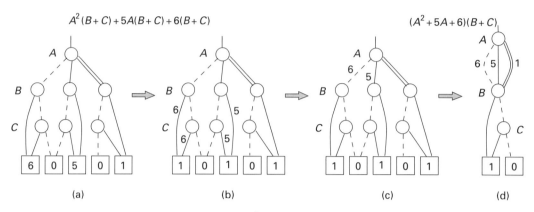

Figure 7.13   Normalization of the TED for $F = (A^2 + 5A + 6)(B + C)$

7.4.5   Canonicity of Taylor expansion diagrams

It now remains to be shown that an ordered, reduced, and normalized Taylor expansion diagram is canonical; i.e., for a fixed ordering of variables, any algebraic expression is represented by a unique reduced, ordered, and normalized TED. First, we recall Taylor's theorem, proved in [68].

THEOREM 7.8 *(Taylor's theorem [68]) Let f(x) be a polynomial function in the domain R, and let $x = x_0$ be any point in R. There exists one and only one unique Taylor series with center x_0 that represents f(x) according to Eq. 7.17.*

This theorem states the uniqueness of the Taylor series representation of a function, evaluated at a particular point (in our case at $x = 0$). This is a direct consequence of the fact that the successive derivatives of a function evaluated at a point are unique. Using Taylor's theorem and the properties of reduced and normalized TEDs, it can be shown that an ordered, reduced, and normalized TED is canonical.

THEOREM 7.9 *For any multivariate polynomial f with integer coefficients, there is a unique (up to isomorphism) ordered, reduced, and normalized Taylor expansion diagram denoting f, and any other Taylor expansion diagram for f contains more vertices. In other words, an ordered, reduced, and normalized TED is minimal and canonical.*

PROOF   The proof of this theorem follows directly the arguments used to prove the canonicity and minimality of BDDs [1] and *BMDs. [2]

Uniqueness First, a reduced TED has no trivial redundancies; the redundant nodes are eliminated by the reduce operation. Similarly, a reduced TED does not contain any isomorphic sub-graphs. Moreover, after the normalization step, all common sub-expressions are shared by further application of the reduce operation. By virtue of Taylor's theorem, all the nodes in an ordered, reduced, and normalized TED are unique and distinguished.

Canonicity We now show that the individual Taylor expansion terms, evaluated recursively, are uniquely represented by the internal nodes of the TED. First, for polynomial functions, the Taylor series expansion at a given point is finite and, according to Taylor's theorem, the series is unique. Moreover, each term in the Taylor series corresponds to the successive derivatives of the function evaluated at that point. By definition, the derivative of a differentiable function evaluated at a particular point is also unique. Since the nodes in the TED correspond to the recursively computed derivatives, every node in the diagram uniquely represents the function computed at that node. Since every node in an ordered, reduced, and normalized TED is distinguished and uniquely represents a function, the Taylor expansion diagram is canonical.

Minimality We now show that a reduced, ordered, and normalized TED is also minimal. This can be proved by contradiction. Let G be a graph corresponding to a reduced, normalized, and, hence, canonical TED representation of a function f. Assume that there exists another graph G', with the same variable order as G, representing f that is smaller than G. This would imply that graph G could be reduced to G' by the application of reduce and normalize operations. However, this is not possible, as G is a reduced and normalized representation and contains no redundancies. The sharing of identical terms across different decomposition levels in the graph G has been captured by the reduction operation. Thus G' cannot have a representation for f with fewer nodes than G. Hence G is a minimal and canonical representation for f.

7.4.6   Complexity of Taylor expansion diagrams

Let us now analyze the worst-case size complexity of an ordered and reduced Taylor expansion diagram. For a polynomial function of degree k, decomposition with respect to a variable can produce $k+1$ *distinct* Taylor expansion terms in the worst case.

THEOREM 7.10 *Let f be a polynomial in n variables and maximum degree k. In the worst case, the ordered, reduced, normalized Taylor expansion diagram for f requires* $O(k^{n-1})$ *nodes and* $O(k^n)$ *edges.*

PROOF The top level contains only one node, corresponding to the first variable. Since its maximum degree is k, the number of distinct child nodes at the second level is bounded by $k+1$. Similarly, each of the nodes at this level produces up to $k+1$ child nodes at the next level, giving rise to $(k+1)^2$ nodes, and so on. In the worst case, the number of children increases in geometric progression, with the level i containing up to $(k+1)^{i-1}$ nodes. For an n-variable function, there will be $n-1$ such levels, with the nth level containing just two terminal nodes, 1 and 0. Hence the total number of internal nodes in the graph is $N = \sum_{i=0}^{n-1} (k+1)^i = \frac{(k+1)^n - 1}{k}$. The number of edges E can be similarly computed as $E = \sum_{i=1}^{n} (k+1)^i = \frac{(k+1)^{n+1} - 1}{k} - 1$, since there may be up to

$(k+1)^n$ terminal edges leading to the 0 and 1 nodes. Thus, in the worst case, the total number of internal nodes required to represent an n-variable polynomial with degree k is $O(k^{n-1})$ and the number of edges is $O(k^n)$.

One should keep in mind, however, that the TED variables represent symbolic, word-level signals, and the number of such signals in the design is significantly smaller than the number of bits in the bit-level representation. Subsequently, even an exponential size of the polynomial with a relatively small number of such variables may be acceptable. Moreover, for many practical designs the complexity is not exponential.

Finally, let us consider the TED representation for functions with variables encoded as n-bit vectors, $X = \sum_{i=0}^{n-1} 2^i x_i$. For linear expressions, the space complexity of TED is linear in the number of bits n, the same as for *BMDs. For polynomials of degree $k \geq 2$, such as X^2, etc., the size of the *BMD representation grows polynomially with the number of bits, as $O(n^k)$. For K*BMDs the representation also becomes non-linear, with complexity $O(n^k - 1)$, for polynomials of degree $k \geq 3$. However, for ordered, reduced, and normalized TEDs, the graph remains linear as the number of bits, namely $O(n \cdot k)$, for any degree k, as stated in the following theorem.

THEOREM 7.11 *Consider variable X encoded as an n-bit vector, $X = \sum_{i=0}^{n-1} 2^i x_i$. The number of internal TED nodes required to represent X^k in terms of bits x_i is $k(n-1) + 1$.*

PROOF We shall first illustrate it for the quadratic case, $k = 2$. Let W_n be an n-bit representation of $X : X = W_n = \sum_{i=0}^{n-1} 2^i x_i = 2^{(n-1)} x_{n-1} + W_{n-1}$ where $W_{n-1} = \sum_{i=0}^{n-2} 2^i x_i$ is the part of X containing the lower $(n-1)$ bits. With that,

$$W_n^2 = (2^{n-1} x_{n-1} + W_{n-1})^2 = 2^{2(n-1)} x_{n-1}^2 + 2^n x_{n-1} W_{n-1} + W_{n-1}^2. \tag{7.19}$$

Furthermore, let

$$W_{n-1} = (2^{n-2} x_{n-2} + W_{n-2}), \tag{7.20}$$

and

$$W_{n-1}^2 = (2^{2(n-2)} x_{n-2}^2 + 2^{n-1} x_{n-2} W_{n-2} + W_{n-2}^2). \tag{7.21}$$

Notice that the constant term (0 edge) of W_{n-1} w.r.t. variable x_{n-2} contains the term W_{n-2}, while the linear term (1 edge) of W_{n-1}^2 contains $2^{n-1} W_{n-2}$. This means that the term W_{n-2} can be *shared* at this decomposition level by two different parents. As a result, there are exactly two non-constant terms, W_{n-2} and W_{n-2}^2, at this level, as shown in Fig. 7.14.

In general, at any level l, associated with variable x_{n-l}, the expansion of terms W_{n-l}^2 and W_{n-l} will create *exactly two* different non-constant terms, one representing W_{n-l-1}^2 and the other W_{n-l-1}; plus a constant term 2^{n-l}. The term W_{n-l-1} will be shared, with different multiplicative constants, by W_{n-l}^2 and W_{n-l}.

This reasoning can be readily generalized to arbitrary integer degree k; at each level there will always be exactly k different non-constant terms. Since on the top-variable

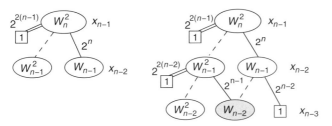

Figure 7.14   Construction of TED for X^2 with n bits

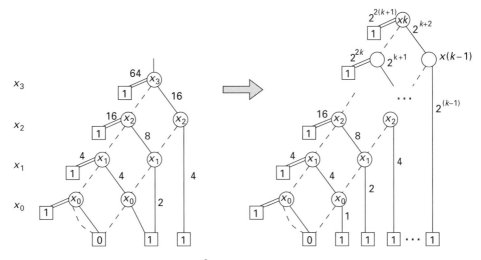

Figure 7.15   Derivation of TED representation for X^2 with n bits

(x_{n-1}) level there is only one node (the root), and there are exactly k non-constant nodes at each of the remaining $(n-1)$ levels, the total number of nodes is equal to $k(n-1)+1$.

The derivation of TED representation for X^2 generalized to n bits is shown in Fig. 7.15.

Table 7.1 compares the worst-case size complexity of the canonical "decision" diagrams described in this chapter in terms of the number of nodes as a function of the size of their operands (bit width n). It shows a significantly lower worst-case complexity for TED than for other representations.

7.4.7   Composition of Taylor expansion diagrams

Taylor expansion diagrams can be composed to compute complex expressions from simpler ones. This section describes general composition rules to compute a new TED as an algebraic sum ($+$) or product ($\cdot$) of two TEDs. The general composition process for TEDs is similar to that of the APPLY operator for BDDs, [1] in the sense that the

Table 7.1 Size complexity of different canonical diagrams

Diagram type	X	$X + Y$	$X \cdot Y$	Xk	c^X
MTBDD	exp.	exp.	exp.	exp.	exp.
EVBDD	lin.	lin.	exp.	exp.	exp.
*BMD	lin.	lin.	lin.	n^k	lin.
K*BMD	lin.	lin.	lin.	n^{k-1}	lin.
TED	const.	const.	const.	$(n-1)k$	–

operations are recursively applied on respective graphs. However, the composition rules for TEDs are specific to the rules of the algebra $(R, \cdot, +)$.

Starting from the roots of the two TEDs, the TED of the result is constructed by recursively constructing all the non-zero terms from the two functions, and combining them, according to the given operation, to form the diagram for the new function. To ensure that the newly generated nodes are unique and minimal, the REDUCE operator is applied to remove any redundancies in the graph.

Let u and v be two nodes to be composed, resulting in a new node q. Let $var(u) = x$, and $var(v) = y$ denote the decomposing variables associated with the two nodes. The top node q of the resulting TED is associated with the variable with the higher order, i.e., $var(q) = x$, if $x \geq y$, and $var(q) = y$ otherwise. Let f, g be two functions rooted at nodes u, v, respectively, and h be a function rooted at the new node q.

For the purpose of illustration, we describe the operations on linear expressions, but the analysis is equally applicable to polynomials of arbitrary degree. In constructing these basic operators, we must consider several cases:

1. Both nodes u, v are terminal nodes. In this case a new *terminal* node q is created as $val(q) = val(u) + val(v)$ for the ADD operation, and as $val(q) = val(u) \cdot val(v)$ for the MULT operation.

2. At least one of the nodes is non-terminal. In this case, the TED construction proceeds according to the variable order. Two cases need to be considered here: (a) when the top nodes u, v have the same index, and (b) when they have different indices. The detailed analysis of both cases is given in [70]. Here, we show the multiplication of two diagrams rooted at variables u and v with the same index.

$$\begin{aligned} h(x) &= f(x) \cdot g(x) \\ &= (f(0) + xf'(0)) \cdot (g(0) + xg'(0)) \\ &= [f(0)g(0)] + x[f(0)g'(0) + f'(0)g(0)] + x^2[f'(0)g'(0)]. \end{aligned} \qquad (7.22)$$

In this case, the 0-child of q is obtained by pairing the 0-children of u, v. Its 1-child is created as a sum of two cross products of 0- and 1-children, thus requiring an additional ADD operation. Also, an additional 2-child (representing the quadratic term) is created by pairing the 1-children of u, v. This is shown in Fig. 7.16.

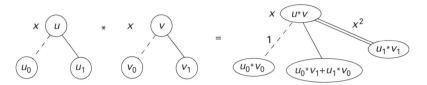

Figure 7.16   Multiplicative composition for nodes with the same variables

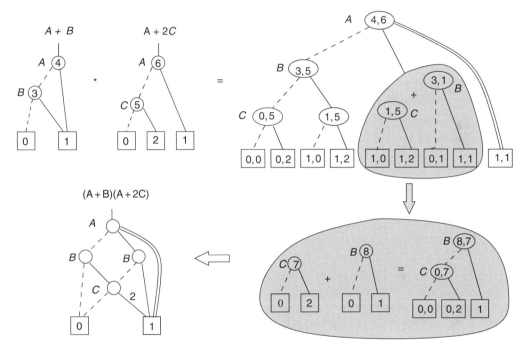

Figure 7.17   Example of MULT composition: $(A+B)(A+2C)$

Figure 7.17 illustrates the application of the ADD and MULT procedures to two TEDs. As shown in the figure, the root nodes of the two TEDs have the same variable index. The MULT operation requires the following steps: (i) performing the multiplication of the respective constant (0-) and linear (1-) children nodes; and (ii) generating the sum of the cross-products of the 0- and 1-children. On the other hand, the two TEDs corresponding to the resulting cross product, as highlighted in the figure, have different variable indices for their root nodes. In this case, the node with the lower index corresponding to variable C is added to the 0-child of the node corresponding to variable B.

It should be noted that the ADD and MULT procedures described above will initially produce non-normalized TEDs, with numeric values residing only in the terminal nodes, requiring further normalization. When these operations are performed on normalized TEDs, with weights assigned to the edges, the following modification is required: when the variable indices of the root nodes of f and g are different, the edge weights have to be propagated down to the children nodes recursively. Downward propagation of edge weights results in dynamic updating of the edge weights of the children nodes. In each recursion step, this propagation of edge weights down to the children proceeds until the weights reach the terminal nodes. The numeric values are updated only in the terminal nodes. Every time a new node is created, the REDUCE and NORMALIZE operations must be performed to remove any redundancies from the graph and generate a minimal and canonical representation.

7.4.8   Design modeling and verfication with TEDs

Using the operations described in the previous section, Taylor expansion diagrams can be constructed to represent various computations over integers in a compact, canonical form. The compositional operators ADD and MULT can be used to compute any combination of arithmetic functions by operating directly on their TEDs. However, the representation of Boolean logic, often present in RTL designs, requires special attention since the output of a logic block must evaluate to a Boolean rather than to an integer value.

Boolean logic

As with *BMDs, one can also define TED operators for Boolean logic, OR, AND, and XOR, where both the range and domain of function are Boolean. This can be done in much the same way as for *BMDs. In fact, the TED and *BMD for a Boolean logic are identical, because they require only the first moment decomposition (refer to Fig. 7.5).

Similarly, one can derive other operators that rely on Boolean variables as one of their inputs, with other inputs being word-level. One such example is the multiplexer, MUX, $(c, X, Y) = c \cdot X + (1 - c)$, where c is a binary control signal, and X and Y are word-level inputs.

In general, TED, which represents an integer-valued function, will also correctly model designs with arithmetic and Boolean functions. Note that the ADD $(+)$ function will always create correct integer results over Boolean and integer domains, because Boolean variables are treated as binary $(0,1)$, a special case of integer. However, the MULT $(\cdot)$ function may create powers of Boolean variables, x^k, which should be reduced to x. A minor modification of TED is made to account for this effect, so that the Boolean nature of variable x can be maintained in the representation. Such modified Taylor expansion diagrams are also canonical.

TED construction for RTL designs

The TED construction for an RTL design starts with building trivial TEDs for its primary inputs. Partial expansion of the word-level input signals is often necessary when one or more bits from any of the input signals fan out to other parts of the design. This is the case in the designs shown in Figs. 7.18 (a) and (b), where bits $a_k = A[k]$ and $b_k = B[k]$ are derived from word-level variables A and B. In this case, the word-level variables must be decomposed into several word-level variables with shorter bit-widths. In our case, $A = 2^{(k+1)}A_{\text{hi}} + 2^k a_k + A_{\text{lo}}$ and $B = 2^{(k+1)}B_{\text{hi}} + 2^k b_k + B_{\text{lo}}$, where $A_{\text{hi}} = A[n-1:k+1]$, $a_k = A[k]$, and $A_{\text{lo}} = A[k-1:0]$, and similarly for variable B.

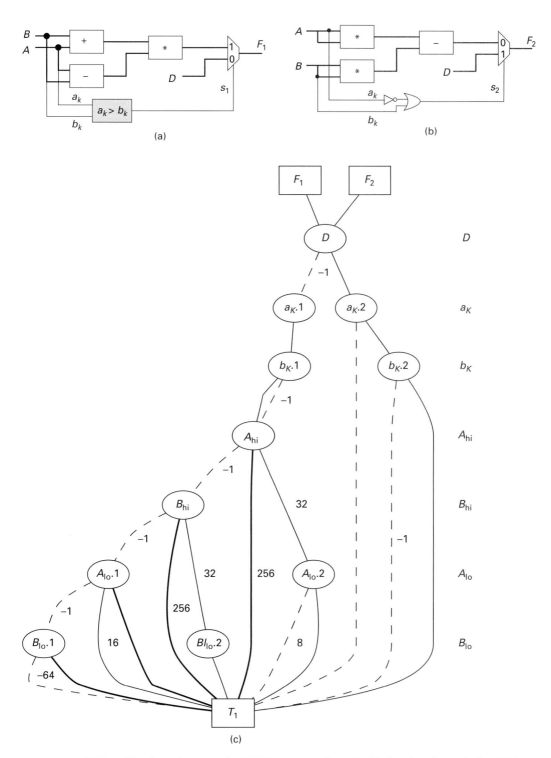

Figure 7.18   RTL verification using canonical TED representation: (a), (b) functionally equivalent RTL
modules; (c) the isomorphic TED for the two designs

Variables $A_{hi}, a_k, A_{lo}, B_{hi}, b_k, B_{lo}$ form the *abstracted* primary inputs of the system. The basic TEDs are readily generated for these abstracted inputs from their respective bases (A_{hi}, a_k, A_{lo}) and (B_{hi}, b_k, B_{lo}).

Once all the abstracted primary inputs are represented by their TEDs, Taylor expansion diagrams can be constructed for all the components of the design. Taylor expansion diagrams for the primary outputs are then generated by systematically composing the constituent TEDs in topological order, from the primary inputs to the primary outputs. For example, to compute $A + B$ in Figs. 7.18(a) and (b), the ADD operator is applied to functions A and B (each represented in terms of their abstracted components). The subtract operation, $A - B$, is computed by first multiplying B with a constant -1 and adding the result to the TED of A. The multipliers are constructed from their respective inputs using the MULT operator, and so on. To generate a TED for the output of the multipliers, the Boolean functions s_1 and s_2 first need to be constructed as TEDs. Function s_1 is computed by transforming the single-bit comparator $a_k > b_k$ into a Boolean function and is expressed as an algebraic equation, $s_1 = a_k \wedge \bar{b}_k = a_k \cdot (1 - b_k)$, as described in this section. Similarly, $s_2 = \bar{a}_k \vee b_k$ is computed as $s_2 = 1 - a_k \cdot (1 - b_k)$ and represented as a TED. Finally, the TEDs for the primary outputs are generated using the MUX operator with the respective inputs. As a result of such a series of composition operations, the outputs of the TED represent multi-variate polynomials in terms of the primary inputs of the design.

TED-based verification

After having constructed the respective ordered, reduced, and normalized Taylor expansion diagram for each design, the test for functional equivalence is performed by checking for isomorphism of the resulting graphs. In fact, the TED-based verification is similar to that using BDDs and BMDs: the generation of the TEDs for the two designs under verification takes place in the same TED manager; when the two functions are equivalent, both functions point to the same root of the common TED. This is shown in Fig. 7.18(c).

It should be noted that the arithmetic operations in these designs assume that no overflow is produced by any of the intermediate signal. That is, functions F_1 and F_2 are functionally equivalent under the *infinite-precision computation* model. This limitaion is a natural consequence of the design representation on the abstract level, where a notion of the individual bits is not available.

7.4.9   Implementation and experimental results

A prototype version of TED software for behavioral HDL designs has been implemented using as a front-end a popular high-level synthesis system, GAUT [71]. This system was selected for its commercial quality, robustness, and open architecture. The input to the system is a behavioral VHDL or C description of the design. The design is parsed and the extracted dataflow is automatically transformed into canonical TED representation.

The core computational platform of the TED package consists of a *manager* that performs the construction and manipulation of the graph. It provides routines to store and manipulate the nodes, edges, and terminal values uniquely, in order to keep the diagrams canonical. To support canonicity, the nodes are stored in a hash table, implemented as *unique table*, similar to that of the CUDD package. [14, 15] The table contains a *key* for each vertex of the TED, computed from the node index and the attributes of its children and the edge weights. As a result, the equivalence test between two TEDs reduces to a simple scalar test between the identifiers of the corresponding vertices.

Variable ordering

Since TEDs are a canonical representation subject to the imposition of a total ordering of the variables, it is desirable to search for a variable order that would minimize the size of TEDs. Dynamic variable ordering for TEDs is based on local swapping of adjacent variables in the diagram, similar to those employed in BDD ordering. [72, 73] It has been shown that, as with BDDs, local swapping of adjacent variables does not affect the structure of the diagram outside the swapping area.

In addition, TEDs can be subjected to static ordering. Typically, the variables are ordered topologically, from primary inputs to primary outputs, in the order in which the signals appear in the design specification. Coefficients are usually represented as weights associated with TED edges. In some cases, however, it may be beneficial to treat some of the coefficients as special variables, rather than weights associated with edges, and place them in the TED graph above all the signal variables. This is particularly important when TEDs are used for the purpose of expression simplification and TED decomposition, as it facilitates symbolic factorization and common subexpression elimination. [74]

Experimental set-up

Several experiments were performed using the prototype TEDify software on a number of dataflow designs described in behavioral VHDL. The designs range from simple algebraic (polynomial) computations to those encountered in signal- and image-processing algorithms. Simple RTL designs with Boolean-algebraic interface were also tested.

The experiments with TED were conducted as follows. The design described in behavioral VHDL or C was parsed by a high-level synthesis system GAUT. [71] The extracted dataflow was then automatically translated into a canonical TED representation using the experimental software TEDify. Comparisons against *BMDs were conducted to demonstrate the power of abstraction of TED representation. For this purpose, each design was synthesized into a structural netlist from which *BMDs were constructed. In most cases, BDDs could not be constructed, owing to their prohibitive size, and they are not reported. Experiments confirm that word-size abstraction by TEDs results in much smaller graph size and computation times than for *BMDs.

Table 7.2 Size of TED vs. Boolean logic for the design in Fig. 7.18a

Bits	*BMD		TED	
(k)	Size	CPU time	Size	CPU
4	4620	107 s	194	44 s
8	15 k	87 s	998	74 s
12	19 k	93 s	999	92 s
16	23.9 k	249 s	4454	104 s
18	Timeout	>12 h	12.8 k	29 min
20	Timeout	>12 h	Timeout	>12 h

Verification of high-level transformations

During the process of architectural synthesis, the initial HDL description often proceeds through a series of high-level transformations. For example, the computation
$AC + BC$ can be transformed into an equivalent one, $(A + B)C$, which better utilizes
the hardware resources. Taylor expansion diagrams are ideally suited to verifying the
correctness of such transformations by proving the equivalence of the two expressions,
regardless of the word size of the input or output signals. We performed numerous
experiments to verify the equivalence of such algebraic expressions. Results indicate
that both time and memory usage required by TEDs is orders of magnitude smaller
than with *BMDs. For example, the expression $(A + B)(C + D)$, where A, B, C, and D
are n-bit vectors, has a TED representation containing just four internal nodes,
regardless of the word size. The size of the *BMD for this expression varies from 418
nodes for the 8-bit vectors, to 2808 nodes for 32-bit variables. Binary decision diagram
graphs could not be constructed for operands with more than 15 bits.

RTL verification

As mentioned earlier, TEDs offer the flexibility of representing designs containing both
arithmetic operators and Boolean logic. We used the generic designs of Figure 7.18 and
performed a set of experiments to observe the efficiency of TED representation under
varying size of Boolean logic. The size of the algebraic signals A, B was kept constant at
32 bits, while the word size of the comparator (or the equivalent Boolean logic) was varied
from 1 to 20. As the size of Boolean logic present in the design increases, the number of
bits extracted from A, B also increases (the figure shows it for single bits). Table 7.2 gives
the results obtained with TED and compares them with those of *BMDs. Note that, as the
size of Boolean logic increases, the TED size converges to that of *BMD. This is to be
expected, as *BMDs can be considered as a special (Boolean) case of TEDs.

Array processing

An experiment was also performed to analyze the capability of TEDs of representing
computations performed by an array of processors. The design that was analyzed is an

Table 7.3 PE computation: $(A_i^2 - B_j^2)$

Array size	*BMD		TED	
$(n \times n)$	Size	CPU time	Size	CPU time
4×4	123	3 s	10	1.2 s
6×6	986	3.4 s	14	1.5 s
8×8	6842	112 s	18	1.6 s
16×16	Out of memory	–	34	8.8 s

$n \times n$ array of configurable processing elements (PE), which is a part of a low-power motion-estimation architecture. [75] Each processing element can perform two types of computations on a pair of 8-bit vectors, A_i, B_i, namely $(A_i - B_j)$ or $(A_i^2 - B_j^2)$, and the final result of all PEs is then added together. The size of the array was varied from 4×4 to 16×16, and the TED for the final result was constructed for each configuration.

When the PEs are configured to perform subtraction $(A_i - B_j)$, both TEDs and *BMDs can be constructed for the design. However, when the PEs are configured to compute $A_i^2 - B_j^2$, the size of *BMDs grows quadratically. As a result, we were unable to construct *BMDs for the 16×16 array of 8-bit processors. In contrast, the TEDs were constructed easily for all the cases. The results are shown in Table 7.3. Note that we were unable to construct the BDDs for any size n of the array for the quadratic computation.

DSP computations

One of the most suitable applications for TED representation is the algorithmic description of dataflow computations, such as digital signal and image processing algorithms. For this reason, we have experimented with the designs that implement various DSP algorithms.

Table 7.4 presents some data related to the complexity of the TEDs constructed for these designs. The first column in the table describes the computation implemented by the design. These include: *FIR* and *IIR* filters, fast Fourier transform (*FFT*), elliptical wave filter (*Elliptic*), least mean square computation (*LMS*128), discrete cosine transform (*DCT*), matrix product computation (*ProdMat*), and Kalman filter (*Kalman*). Most of these designs perform algebraic computations by operating on vectors of data, which can be of arbitrary size. The next column gives the number of inputs for each design. While each input is a 16-bit vector, TED represents them as word-level symbolic variables. Similarly, the next column depicts the number of 16-bit outputs. The remaining columns of the table show: the BMD size (number of nodes); the CPU time required to construct the BMD for the 16-bit output words; the TED size (number of nodes) required to represent the entire design. The CPU time required to generate TED diagrams does not account for the parsing time of the GAUT front end.

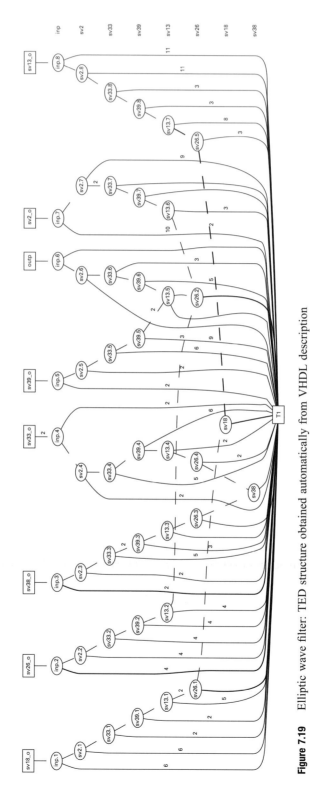

Figure 7.19 Elliptic wave filter: TED structure obtained automatically from VHDL description

Table 7.4 Signal processing applications

Design	Input size	Output size	*BMD size (nodes)	BMD CPU time (s)	TED size (nodes)	TED CPU time (s)
Dup-real	3×16	1×16	92	10	5	1
IIR	5×16	1×16	162	13	7	1
FIR16	16×16	1×16	450	25	18	1
FFT	10×16	8×16	995	31	29	1
Elliptic	8×16	8×16	922	19	47	1
LMS128	50×16	1×16	8194	128	52	1
DCT	32×16	16×16	2562	77	82	1
ProdMat	32×16	16×16	2786	51	89	1
Kalman	77×16	4×16	4866	109	98	1

Figure 7.19 depicts a multiple-output TED for the elliptical wave filter (design *elliptic*), where each root node corresponds to an output of the design.

Algorithmic verification

This final set of experiments demonstrates the natural capability of Taylor expansion diagrams of verifying the equivalence of designs described at the *algorithmic* level. Consider two dataflow designs computing convolution of two real vectors, $A(i), B(i), i = 0, \ldots N-1$, shown in Fig. 7.20. The design in Fig. 7.20(a) computes the FFT of each vector, computes the product of the FFT results, and performs the inverse FFT operation, producing output *IFFT*. The operation shown in Fig. 7.20(b) computes the convolution directly from the two inputs, $C(i) = \sum_{k=0}^{N-1} A(k) \cdot B(i-k)$. A TED was used to represent these two computations for $N = 4$ and to prove that they are, indeed, equivalent. Figure 7.21 depicts the TED for vector C of the convolution operation, isomorphic with the vector *IFFT*. All graphs are automatically generated by our TED-based verification software.

As illustrated by the above example, TEDs can be suitably augmented to represent computations in the complex domain. In fact, it can be shown that TEDs can represent polynomial functions over an *arbitrary field*. The only modification required is that the weights on the graph edges are elements of the field, and that the composition MULT and ADD are performed with the respective operators of the field. Subsequently, TEDs can also be used to represent computations in the *Galois field*. [76]

7.4.10   Limitations of TED representation

Taylor expansion diagrams have several natural limitations. As mentioned earlier, TEDs can only be used to represent infinite-precision arithmetic, and cannot represent modular arithmetic. Furthermore, they can only represent functions that have *finite* Taylor expansions, and, in particular, multi-variate polynomials with finite-integer degrees. For polynomials of finite-integer degree $k \geq 1$, successive differentiation of

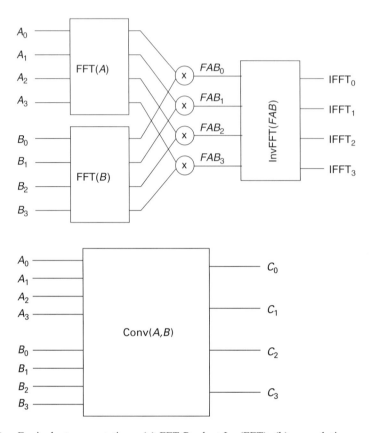

Figure 7.20   Equivalent computations: (a) FFT-Product-Inv(FFT); (b) convolution

the function ultimately leads to zero, resulting in a finite number of terms. However, those functions that have infinite Taylor series (such as a^x, where a is a constant) cannot be represented with a finite TED graph. To represent exponentials using TEDs, one must expand the integer variable into bits, $X = \{x_{n-1}, x_{n-2}, \ldots, x_0\}$, and use the TED formulas to represent the function in terms of the bits. Such a TED would be structurally similar to the *BMD representation of the function.

While TED representation naturally applies to functions that can be modeled as *finite polynomials*, the efficiency of TEDs relies on their ability to encode the design in terms of its *word-level* symbolic inputs, rather than bit-level signals. This is the case with the simple RTL designs shown in Fig. 7.18, where all input variables and internal signals have simple, low-degree polynomial representation. The abstracted word-level inputs of these designs are created by partial bit selection (a_k, b_k) at the primary inputs, and a polynomial function can be constructed for the outputs. However, if any of the internal or output signals is partitioned into sub-vectors, such sub-vectors cannot be represented as polynomials in terms of the symbolic, word-level input variables, but depend on the individual bits of the inputs. The presence of such signal splits creates a fundamental problem for the polynomial representations, and TEDs cannot be used

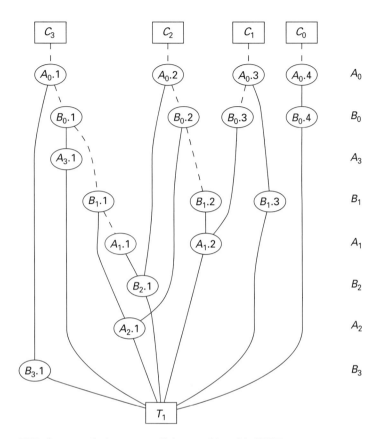

Figure 7.21 TED for convolution vector C, isomorphic with *IFFT*

efficiently in those cases. For similar reasons, TED cannot represent modular arithmetic. An attempt to fix this problem was proposed in [76], by modeling the discrete functions as finite, word-level polynomials in the Galois field (GF). The resulting polynomials, however, tend to be of much higher degree than the original function, with the degree depending on the signal bit width, making the representation less efficient for practical applications. This is the case where TEDs can exhibit space explosion similar to that encountered in BDDs and BMDs.

Another limitation of TEDs is that they cannot represent relational operators (such as comparators, $A \geq B$, A $==$ B, etc.) in symbolic form. This is because the Taylor series expansion is defined for functions and not for relations. Relations are characterized by discontinuities over their domain and are not differentiable. To use TEDs to represent relational operators, often encountered in RTL descriptions, the expansion of word-level variables and bit vectors into their bit-level components is required.

Despite these limitations, TEDs can be successfully used for verifying equivalence of high-level, behavioral, and algorithmic descriptions. Such algorithmic descriptions typically do not exhibit signal splits, resulting in polynomial functions over word-level variables.

7.4.11   Conclusions and open problems

This section described a compact, canonical, graph-based representation, called the Taylor expansion diagram (TED). It has been shown that, for a fixed ordering of variables, the TED is a canonical representation that can be used to verify equivalence of arithmetic computations in dataflow designs. It has been shown how TEDs can be constructed for behavioral and some RTL design descriptions. The power of abstraction of TEDs makes them particularly applicable to dataflow designs specified at the behavioral and algorithmic level.

For larger systems, especially those involving complex bit-select operations, and containing large portions of Boolean logic, relational operators, and memories, TEDs can be used to represent those portions of the design that can be modeled as polynomials. Equivalence checking of such complex design typically involves finding structurally similar points of the designs under verification. The TED data structure can be used here to raise the level of abstraction of large portions of designs, aiding in the identification of such similar points and in the overall verification process. In this sense, TEDs complement existing representations, such as BDDs and *BMDs, in places where the level of abstraction can be raised.

The experiments demonstrate the applicability of TED representation to verification of dataflow designs specified at behavioral and algorithmic levels. This includes portions of algorithm-dominant designs, such as signal processing for multimedia applications and embedded systems. Computations performed by those designs can often be expressed as polynomials and can be readily represented with TEDs. The test for functional equivalence is then performed by checking the isomorphism of the resulting graphs. Of particular promise is the use of TEDs in the verification of algorithmic descriptions, where the use of symbolic, word-level operands, without the need to specify bit width, is justified. A number of open problems remain to be researched to make TEDs a reliable data structure for high-level design representation and verification.

In addition to these verification-related applications, TEDs prove useful in algorithmic and behavioral synthesis and optimization for DSP and dataflow applications. [74] Taylor expansion diagram data structures, representing a functional view of the computation, can serve as an efficient vehicle to obtain a structural representation, namely the dataflow graph (DFG). This can be obtained by means of graph decomposition, which transforms the functional TED representation into a structural DFG representation. By properly guiding the decomposition process, the resulting DFG can provide a better starting point for the ensuing architectural (high-level) synthesis, than that extracted directly from the original HDL specification.

7.5   Representation of multiple-output functions over finite fields

This section presents a method for representing multiple-output, binary, and word-level functions in GF(N) (N $= p^m$; p a prime number and m a non-zero positive integer) based on decision diagrams (DD). The presented DD is canonical and can be made minimal with respect to a given variable order. The DD has been tested on benchmarks including integer multiplier circuits and the results show that it can produce better node compression (more than an order of magnitude in some cases) than shared BDDs. The benchmark results also reflect the effect of varying the input and output field sizes on the number of nodes. Methods of graph-based representation of characteristic and encoded characteristic functions in GF(N) are also presented. The performance of the proposed representations has been studied in terms of average path lengths and the actual evaluation times with 50 000 randomly generated patterns on many benchmark circuits. All these results reflect that the proposed technique can outperform existing techniques.

7.5.1 Previous work

Finite fields have numerous applications in public-key cryptography [78] to encounter channel errors and for protection of information, error control codes, [78] and digital signal processing [79]. Finite fields gained significance with practical lucrativeness of the elliptic-curve crypto systems. The role of finite fields in error control systems is well established and contributes to many fault-tolerant designs. In the EDA industry, the role of multi-valued functions, especially in the form of multi-valued decision diagrams (MDD), is well described in [80,81]. Word-level diagrams can be useful in high-level verification, logic synthesis, [1,82] and software synthesis. [83] Multi-valued functions can also be represented in finite fields, as shown in [84]. Finite fields can represent many arithmetic circuits very efficiently. [85] Also, there are fine-grain FPGA structures for which arithmetic circuits in finite fields seem to be highly efficient. The varied use of finite fields leads to the design of high-speed, low-complexity systolic VLSI realizations. [86] Fast functional simulation in the design cycles is a key step in all these applications. [87]

Most existing techniques for word-level representation, e.g., [2, 88], are not capable of efficiently representing arbitrary combinations of bits or nibbles, i.e., sub-vectors, within a word. The proposed framework for representing circuits can deal with these types of situation by treating each sub-vector as a word-level function in GF(2^m), where m denotes the number of bits within a sub-vector. The word-level functions are then represented as canonic word-level graphs. Hence the proposed technique offers a generalized framework for verifying arbitrary combinations of bits or words.

Another situation where existing word-level techniques seem to have difficulty is representing *non-linear* design blocks, such as comparators, at the register transfer level, RTL (e.g., in the integer domain). The proposed framework does not suffer from this critical shortcoming.

As an example of representing an arbitrary combination of output bits in a multiple-output function, let's consider a four-input eight-output binary function. The MSB expressed by $f = \sum m(10,11,12,13,14,15)$,[1] and the LSB $g = \sum m(4,5,6,7,8,9,10,11, 14,15)$ can be represented on the same diagram as shown in Fig. 7.22. [89] The BDD-based representation of this circuit will require a larger number of nodes.

---

[1] The notation $h = \sum m(p_1, p_2, \ldots, p_q)$ is used to represent the truth-table of a function where each p_r ($1 \le r \le q$) is the decimal equivalent of a row in the input part of the table with an output of 1, i.e., each p_r is a *minterm* from the ON set in its decimal form.

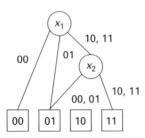

Figure 7.22   Representing two bits simultaneously

Although research has been done on representing circuits in finite fields, [85, 90] the theoretical basis was carried out, for example, in the spectral domain for a fixed value, e.g., four in [90]. Unlike these techniques, this section presents a generalized framework for the design, verification, and simulation of circuits in finite fields based on the MDD-like graph-based form. The proposed DD has advantages over other diagrams, such as [2], in that, in addition to applications in multiple-valued algebra, it is not restricted to word boundaries, but instead it can be used to represent and verify any combination of output bits. Unlike [91], which is not a DD and, hence, lacks many features present in a DD, the proposed diagram does not have such shortcomings. Also, unlike [81], the proposed DD represents finite fields and extension fields, while [81] is based on MIN-MAX post-algebra.

Owing to its canonicity, the proposed technique can be used for verifying circuits at the bit or word-level by checking for graph isomorphism, which can be done very quickly.

Fast evaluation times of multiple-output functions is significant in the areas of logic simulation, testing, satisfiability, and safety checking. [92, 93] The proposed DDs also offer much shorter average path lengths and, hence, evaluation times [94] than shared BDDs, with a varying trade-off between evaluation times and spatial complexity.

7.5.2   Background and notation

Finite fields

Let GF(N) denote a set of N elements, where $N = p^m$, p is a prime number and m a non-zero positive integer, with two special elements 0 and 1 representing the additive and multiplicative identities respectively, and two operators; addition '$+$' and multiplication '$\cdot$'. The function GF(N) defines a finite field, also known as a *Galois field*, if it forms a *commutative ring* with identity over these two operators, in which every element has a multiplicative inverse. In other words, GF(N) defines a finite field if the following properties hold:

- *Associativity*: $\forall a,b,c \in$ GF(N) $(a+b)+c=a+(b+c)$, and $(a \cdot b) \cdot c = a \cdot (b \cdot c)$.
- *Identity*: $\forall a \in$ GF(N) $a+0=0+a=a$, and $a \cdot 1 = 1 \cdot a = a$. Here, 0 and '1' are the additive and multiplicative identities respectively.
- *Inverse*: $\forall a \in$ GF(N) $\exists -a, a^{-1} \in$ GF(N), such that $a+(-a)=0$ and $a \cdot a^{-1}=1$. Here, $-a$ and a^{-1} are the additive and multiplicative inverses respectively.

Table 7.5 Generation of GF(2^3)

Exponential representation		Polynomial representation		Bit vector
0	$=$	0	$\leftrightarrow$	[0,0,0]
a^0	$=$	1	$\leftrightarrow$	[0,0,1]
a^1	$=$	a^2	$\leftrightarrow$	[0,1,0]
a^2	$=$	a^2	$\leftrightarrow$	[1,0,0]
a^3	$=$	$a+1$	$\leftrightarrow$	[0,1,1]
a^4	$=$	a^2+a	$\leftrightarrow$	[1,1,0]
a^5	$=$	$a^3+a^2=a^2+a+1$	$\leftrightarrow$	[1,1,1]
a^6	$=$	a^2+1	$\leftrightarrow$	[1,0,1]

- *Commutative*: $\forall a,b \in$ GF(N) $a+b=b+a$, and $\forall c,d \in$ GF(N)-$\{0\}c \cdot d=d \cdot c$.
- *Distributive*: '$\cdot$' distributes over '$+$', i.e. $\forall a,b,c \in$ GF(N) $a \cdot (b+c)=(a \cdot b)+(a \cdot c)$.

Here, p, which is a prime number, is called the characteristic of the field, and satisfies the following conditions:

$$(a)\ \underbrace{1+1+\cdots+1}_{p \text{ times}}=0.$$

$$(b)\ pa=0,\ \forall\ a \in \text{GF}(N).$$

Also $\forall a \in$ GF(N), $a^N=a$, and for $a \neq 0$, $a^{N-1}=1$. The elements of GF(N) can be represented as polynomials over GF(p) of degree, at most, $n-1$. There exists an element $a \in$ GF(N) for which the powers of $a, a^2, \ldots, a^{N-1}$ are distinct and represent the non-zero elements of the field. Here a is called the *primitive element* of the field. Additional properties of GF(N) can be found in [78,84].

Generation of finite fields

A polynomial $p(x)$ over GF(p^m) is said to be primitive if it is *irreducible* (i.e., cannot be factored into lower degree polynomials), and if the smallest positive integer r for which $p(x)$ divides x^r-1 is $r=p^m-1$.

For example, the polynomial $p(x)=x^3+x+1$ is primitive over GF(2), because the smallest positive integer for which it is a divisor of x^r-1 is $r=7=2^3-1$, i.e., x^7-1.

Finite fields over GF(2^m) and $m \geq 2$ can be generated with primitive polynomials (PP) of the form $p(x)=x^m+\sum_{i=0}^{m-1}c_ix^i$, where $c_i \in$ GF(2). [78]

For example, given the PP $p(x)=x^3+x+1$, we can generate GF(8) as follows. Let a be a root of $p(x)$, i.e., $p(a)=0$. Hence, $a^3=a+1=0$ or $a^3=a+1$. In general, any element $\beta \in$ GF(2^m) can be represented in this *polynomial form* as $\beta(x)=\sum_{i=0}^{m-1}\beta_ix^i$, where $\beta_i \in \{0,1\}$. In this way, all the elements of GF(8) can be generated as shown in Table 7.5. Note that since each coefficient $\beta_i \in \{0,1\}$, each element in its polynomial form can also be represented as a bit vector, as shown in the third column. The bit vectors can be stored as integers in a computer program. For example, the element a^4 is 5 in decimal, and can be stored as an integer.

+	0	1	α	β
0	0	1	α	β
1	1	0	β	α
α	α	β	0	1
β	β	α	1	0

$\times$	0	1	α	β
0	0	0	0	0
1	0	1	α	β
α	0	α	β	1
β	0	β	1	α

(a) Addition　　　　(b) Multiplication

Figure 7.23　Addition and multiplication over GF(4)

Operations over finite fields

For any $a,\beta \in GF(2^m)$, if a and β are in their polynomial forms as $a(x) = \sum_{i=0}^{m-1} a_i x^i$ and $\beta(x) = \sum_{i=0}^{m-1} \beta_i x^i$, where $a_i, \beta_i \in \{0,1\}$ and $0 \le i < m$, then multiplication over GF (2^m) can be defined as $w(x) = a(x) \cdot \beta(x) \bmod p(x)$, where $p(x)$ represents the PP used to generate the fields [78, 84, 95]. As an example, let $a, \beta \in GF(4)$, which is generated with the PP $p(x) = x^2 + x + 1$. Also let $a(x) = x$ and $\beta(x) = x + 1$. Then

$$
\begin{aligned}
a \cdot \beta &= a(x) \cdot \beta(x) \bmod p(x) \\
&= x \cdot (x + 1) \bmod x^2 + x + 1 \\
&= x^2 + x \bmod x^2 + x + 1 \\
&= 1.
\end{aligned}
$$

If the elements are in their exponential form, as in the first column of Table 7.5, then multiplications can also be carried out as follows. Let $a,b \in GF(2^m)$, $a = a^x$ and $b = a^y$, then $a \cdot b = a^{(x+y) \bmod 2^{m-1}}$, where the addition is carried out over integers. For example, from Table 7.5 $a^5 \cdot a^6 = a^{(5+6) \bmod 7} = a^4$ over GF(8).

Note that $a + \beta$, i.e., addition over $GF(2^m)$, is the bitwise EXOR of the bit vectors corresponding to $a(x)$ and $\beta(x)$. For example, let $a = a^5$ and $b = a^6$ from Table 7.5. We also have $a = [1,1,1]$ and $b = [1,0,1]$. Hence $a + b = [0,1,0]$, which is a.

Figure 7.23 shows multiplication and addition tables over GF(4).

Notation

The following notation is used in this chapter.

Let $I_N = \{0,1, \ldots, N-1\}$, and $\delta: I_N \to GF(N)$ be a one-to-one mapping, with $\delta(0) = 0$. Let $f|_{x_k} = y$, called the *co-factor* of f w.r.t. $x_k = y$, represent the fact that all occurrences of x_k within f are replaced with y, i.e. $f|_{x_k} = y = f(x_1, x_1, \ldots, x_k = y, \ldots, x_n)$. The notation $f|_{x_i} = y_i, x_{i+1} = y_{i+1}, \ldots, x_{i+j} = y_{i+j}$ (or just $f|_{y_i}, y_{i+1}, \ldots, y_{i+j}$ when the context is clear) will be used to represent the replacement of variables $x_i, x_{i+1}, \ldots, x_{i+j}$ with the values $y_i, y_{i+1}, \ldots, y_{i+j}$ respectively.

We shall use the notation $|A|$ to represent the total number of nodes in a graph A.

We have the following in GF(N).

THEOREM 7.12 *A function* $f(x_1, x_2, \ldots, x_k, \ldots, x_n)$ *in GF(N) can be expanded as follows.*

$$
f(x_1, x_2, \ldots, x_k, \ldots, x_n) = \sum_{e=0}^{N-1} g_e(x_k) \ f\ |_{x_k = \delta(e)}, \tag{7.23}
$$

where $g_e(x_k) = 1 - [x_k - \delta(e)]^{N-1}$.

PROOF The proof is made by perfect induction as follows. From the properties of GF (N) we have, for any $a \in$ GF(N) such that $a \neq 0$, $a^{N-1} = 1$. Now, in Theorem 7.12 if $x_k = \delta(r)$, then $g_r(x_k) = 1$ for $r \in I_N$. Furthermore, $g_s(x_k) = 0$ $\forall s \in I_N$ and $s \neq r$. Therefore, only one term, namely $f(x_1, x_2, \ldots, \delta(r), \ldots, x_n)$, remains on the right-hand-side of Eq. 7.20, while all the remaining terms equate to zero. Hence the proof follows.

In Theorem 7.12, $g_e(x_k)$ is called a *multiple-valued literal*.[2] Theorem 7.12 is known as the *literal-based expansion* of functions in GF(N). Theorem 7.12 reduces to the *Shannon's expansion* over GF(2) as follows. If we put $N = 2$ in Eq. 7.20 then

$$f(x_1, x_2, \ldots, x_n) = \sum_{e=0}^{2-1} g_e(x_k) f|_{x_k = \delta(e)}$$

$$= g0(x_k) f|_{x=0} + g_1(x_k) f|_{x=1}$$

$$= \bar{x}_x f|_{x=0} + x_k f|_{x=1},$$

where $\bar{x}$ represents that x appears in its complemented form.

The product of literals is called a *product term* or just a *product*. Two product terms are said to be *disjoint* if their product in GF(N) equates to zero. An expression in GF(N) constituting product terms is said to be disjoint if all of its product terms are pairwise disjoint.

Section 7.5.3 presents the theory behind the graph-based representation and its reduction, with methods for additional node and path optimizations. Section 7.5.9 provides the theory behind representing functions in GF(N) in terms of graph-based characteristic and encoded characteristic functions. The proposed methods offer much shorter evaluation times than existing approaches and Section 7.5.10 provides a technique for calculating the average path lengths for approximating the evaluation times for the proposed representations. The proposed technique has been tested on many benchmark circuits. Finally, in Section 7.5.11, we present the experimental results.

7.5.3   Graph-based representation

Any function in GF(N) can be represented by means of an MDD-like data structure. [81] However, unlike traditional MDDs, which are used to represent functions in the MIN-MAX post-algebra, the algebra of finite fields needs to be considered. Although an MDD type of data structure has been used for representing functions in finite fields in [90], the underlying mathematical framework was considered for GF(4) only, no generalization was proposed for higher-order fields and their extensions, and no experimental results were reported, even though it was reported that generalization can be made. Also, no technique seems to exist, which can further optimize an MDD-like representation of functions in GF(N) by zero-terminal node suppression and

---

[2] The term *literal* was chosen because in GF(2) this expression reduces to the traditional Boolean literal, i.e., it represents a variable or its complement (inverse).

normalization. It should be noted that the technique of [81] has used a type of edge negation based on modular arithmetic. However, modular arithmetic in the form considered in [81] does not naturally comply with extension fields. Since an MDD has been defined in terms of functions in the MIN-MAX post-algebra, to distinguish between these two algebras, the MDD-like representation of functions in finite fields will be called *multiple-output decision diagrams* or MODDs. Hence, traditional MDDs result in a post-algebraic MIN-MAX SOP form, while with the MODD a canonic polynomial expression in GF(N) can be obtained. As an example, the MODD of Fig. 7.27(a), which represents a four-valued function with values in $\{0,1,a,\beta\}$ (assuming $a = 2$ and $\beta = 3$), yields the following expression in the MIN-MAX post-algebra:

$$f(x_1, x_2, x_3) = \beta(x_1^\beta x_2^\beta \vee x_1^{\{1,a\}} x_2^{\{1,a,\beta\}} x_3^\beta \vee x_1^\beta x_2^{\{1,a\}} x_3^\beta \vee x_1^0 x_2^{\{1,a\}} x_3^\beta)$$
$$\vee\, a(x_1^\beta x_2^{\{1,a\}} x_3^a \vee x_1^{\{1,a\}} x_2^{\{1,a,\beta\}} x_3^a \vee x_1^0 x_2^{\{1,a\}} x_3^a)$$
$$\vee\, (x_1^\beta x_2^{\{1,a\}} x_3^1 \vee x_1^{\{1,a\}} x_2^{\{1,a,\beta\}} x_3^1 \vee x_1^0 x_2^{\{1,a\}} x_3^1).$$

Here the symbol $\vee$ has been used to denote MAX. MIN is denoted by the product-like notation. The expression x_i^S, where $\subseteq \{0,1,a,\beta\}$, is a literal defined in the MIN-MAX postalgebra as $x_i^S = MAX\_VALUE$, where $MAX\_VALUE = \beta$ in this case, if $x_i \in \$$; $x_i^S = 0$ otherwise. In contrast $x_i^S = 1$ if $x_i \in s$; $x_i^S = 0$ otherwise in GF(N) (Theorem 7.12). The following multi-variate polynomial results from the MODD in GF(4) by application of Theorem 7.12 followed by expansion and rearranging the terms:

$$f(x_1, x_2, x_3) = \beta x_1^3 x_2^3 + a x_1^2 x_2^3 + x_1 x_2^3 + a x_1^3 x_2^2 + x_1^2 x_2^2 + \beta x_1 x_2^2$$
$$+ x_3 x_2 + \beta x_1^2 x_2 + a x_1 x_2 + \beta x_2^2 x_3 + a x_2 x_3 + \beta x_1^2 x_2^3 x_3$$
$$+ a x_1 x_2^3 x_3 + a x_1^2 x_2^2 x_3 + x_1 x_2^2 x_3 + x_1^2 x_2 x_3 + \beta x_1 x_2 x_3.$$

DEFINITION 7.13 (DECISION DIAGRAM) *A decision diagram in GF(N) is a rooted directed acyclic graph with a set of nodes V containing two types of nodes: (A) A set of N terminal nodes or leaves with out-degree zero, each one labeled with a $\delta(s)$ and $s \in I_N$. Each terminal node u is associated with an attribute value(u) $\in$ GF(N). (b) A set of non-terminal nodes, with out-degree of N. Each non-terminal node v is associated with an attribute var(v) $= x_i$ and $1 \le i \le n$, and another attribute child$_j$(v) $\in$ V, $\forall j \in I_N$, which represents each of the children (direct successors) of v.*

The correspondence between a function in GF(N) and an MODD in GF(N) can be defined as follows.

DEFINITION 7.14 (RECURSIVE EXPANSION) *An MODD in GF(N) rooted at v denotes a function f^v in GF(N) defined recursively as follows: (a) If v is a terminal node, then $f^v = value(v)$, where value(v) $\in$ GF(N). (b) If v is a non-terminal node with var(v) $= x_i$, then f^v is the function*

$$f^v(x_1, x_2, \ldots, x_i, \ldots, x_n) = \sum_{e=0}^{N-1} g_e(x_i) f^{child_e(v)},$$

where $g_e(x_i) = 1 - [x_i - \delta(e)]^{N-1}$.

Each variable x_i and $1 \leq i \leq n$ in an MODD is associated with one or more nodes, which appear at the same *level* in the MODD. More precisely, the nodes associated with variable x_i correspond to level $(i-1)$ and vice versa. Therefore level $-i$, corresponding to variable $x_i + 1$, can contain at most N^i nodes. Hence, the root of the MODD contains exactly one node, and the level before the external nodes can contain, at most, N^{n-2} nodes.

Example 7.2 Let us consider the MODD shown in Fig. 7.24(a). This MODD represents the following function in GF(3): $f(x_1,x_2) = g_1(x_1) + g_2(x_1)g_1(x_2) + a \cdot g_2(x_1)g_2(x_2)$, where $g_r(x_s) = 1 - [x_s \text{ minus; } \delta(r)]^2$.

Here both the levels 0 and 1, corresponding to the variables x_1 and x_2 respectively, contain exactly one node each.

LEMMA 7.15 *Theorem 7.12 results in a disjoint expression, i.e., the product terms in Eq. 7.20 are mutually (pairwise) disjoint.*

PROOF In Eq. 7.20 $g_e(x_k) = 1$ *iff* $x_k = \delta(e)$. For all other values of x_k $g_e(x_k) = 0$. Let us consider any two literals $g_r(x_k)$ and $g_s(x_k)$, such that $r \neq s$. Two cases may arise:

    Case I $x_k \neq \delta(r)$ and $x_k \neq \delta(s)$. In this case both $g_r(x_k)$ and $g_s(x_k)$ will equate to 0. Therefore, $g_r(x_k) \cdot g_s(x_k) = 0$.
    Case II Either $x_k = \delta(r)$ or $x_k = \delta(s)$, but not both. If $x_k = \delta(r)$, then $g_r(x_k) = 1$ and $g_s(x_k) = 0$; otherwise, $g_r(x_k) = 0$ and $g_s(x_k) = 1$. Therefore, $g_r(x_k) \cdot g_s(x_k) = 0$.

Hence the proof follows.

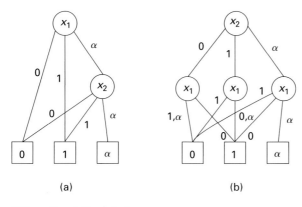

(a)   (b)

Figure 7.24   Effect of variable ordering

Lemma 7.15 yields the following.

THEOREM 7.16 *Each path from the root node to a non-zero terminal node in an MODD represents a disjoint product term in* GF(N).

Example 7.3 Let us consider the MODD in Fig. 7.24(a) representing a function in GF(3). The path $a,1$ represents the product term $X = g_a(x_1)g_1(x_2)$. The path a,a represents the product term $Y = g_a(x_1)g_a(x_2)$. Clearly X and Y are disjoint, because $X \cdot Y = 0$ as $g_1(x_2) \cdot g_a(x_2) = 0$.

7.5.4   Reduction

We have the following from Theorem 7.12.

COROLLARY 7.17 *In Eq. (20), if $\forall \ i,j \in I_N$ and $i \neq j \ f|_{x_k} = \delta(i) = f|_{x_k} = \delta_{(j)}$, then $f = = f|_{x_k} = \delta(0) = f|_{x_k} = \delta(1) = \ldots = f|_{x_k} = \delta(N-1)$.*

PROOF By perfect induction. Let $h = f|_{x_k} = \delta(0) = f|_{x_k} = \delta(1) = \ldots = f|_{x_k} = \delta(N-1)$. Then Eq. 7.20 becomes $f = h \sum_{e=0}^{N-1} g_e(x_k)$.

For any $a \in$ GF(N) such that $a \neq 0$, $a^{N-1} = 1$. Now in Theorem 7.12, if $x_k = \delta(r)$, then $g_r(x_k) = 1$ for $r \in I_N$. Furthermore, $g_s(x_k) = 0$, $\forall s \in I_N$, and $s \neq r$. Therefore, $\sum_{e=0}^{N-1} g_e(x_k)$ becomes 1, which implies $f = h$. Hence the proof.

Based on the above, an MODD can be reduced as outlined in the following.

Reduction rules
There are two reduction rules:

- If all the N children of a node v point to the same node w, then delete v and connect the incoming edge of v to w. This follows from Corollary 7.17.
- Share equivalent sub-graphs.

A DD in GF(N) is said to be ordered if the expansion in Eq. 7.20 is recursively carried out in a certain linear variable order such that on all the paths throughout the graph the variables also respect the same linear order.

A DD in GF(N) said to be reduced *if*: (a) There is no node $u \in V$, such that $\forall i,j \in I_N$ and $i \neq j$, $child_i(u) = child_j(u)$. (b) There are no two distinct nodes $u,v \in V$ which have the same variable names and same children, i.e., $var(u) = var(v)$ and $child_i(u) = child_i(v) \forall i \in I_N$ implies $u = v$. We have the following from the definition of the MODD:

LEMMA 7.18 *For any node v in a reduced DD in* GF(N), *the sub-graph rooted at v is itself reduced.*

Canonicity
A reduced, ordered MODD in GF(N) based on the expansion of Theorem 7.12 is canonical up to isomorphism. This is stated in the following:

THEOREM 7.19 *For any n variable function f(x₁,x₂, . . ., xₙ) in* GF(*N*) *there is exactly one reduced ordered DD in* GF(*N*) *with respect to a specific variable order, which is minimal with respect to the reduction rules.*

PROOF The proof is done by induction on the number of arguments n in f.

Base case: If $n = 0$, then the function yields a constant in GF(N). The resulting reduced, ordered DD in GF(N) has exactly one node, namely the terminal node with a value in GF(N). Therefore, any function in GF(N) with $n = 0$ will have exactly one external node with the same value in GF(N), and, hence, will be unique. Note that any reduced, ordered DD in GF(N) with at least one non-terminal node would yield a non-constant function.

Induction hypothesis Assume that the theorem holds for all functions with $n - 1$ arguments.

Induction step We show that the theorem holds for any function f in GF(N) with n arguments. Without loss of generality, let us assume that the variables are ordered as $(x_1, x_2, \ldots, x_n)$. Considering the first variable x_1,

$$f(x_1, x_2, \ldots, x_n) = \sum_{e=0}^{N-1} g_e(x_1) f|_{x_1} = \delta(e), \qquad (7.24)$$

where $g_e(x_1) = 1 - [x_1 - \delta(e)]^{N-1}$.

Let f^z represent the function realized by the sub-graph rooted by z. In a reduced, ordered DD in GF(N) for f, each $f|_{x_1} = \delta(i)$, $\forall i \in I_N$, is represented by a sub-graph rooted by u_i. Since each $f|_{x_1} = \delta(i)$ is a function of $n - 1$ variables, so each u_i, $\forall i \in I_N$, represents a unique reduced, ordered DD in GF(N), by the induction hypothesis.

Two cases may arise:

Case 1 $\forall i,j \in I_N$ and $i \neq j$, $u_i = u_j$. It must be the case that, $f^{u_i} = f^{u_j}$, which implies that $f|_{x_1} = \delta(i) = f^{u_i} = f^{u_j} = f|_{x_1} = \delta(j)$. Therefore, $u_i = u_j$, $\forall i,j \in I_N$ and $i \neq j$, is a reduced, ordered DD in GF(N) for f. This is also unique, since if it is not, then owing to the ordering x_1, it would appear in the root node v if it appears at all. This would imply that $f = f'$. Therefore, from the definition of a DD in GF(N), we must have $f|_{x_1} = \delta(k) = f^v|_{x_1} = \delta(k) = f^{child_k}(v)$, $\forall k \in I_N$. Since, by assumption, $f|_{x_1} = \delta(i) = f^{u_i} = f^{u_j} = f|_{x_1} = \delta(j)$, $\forall i,j \in I_N$ and $i \neq j$, this would imply that all the children of v would be the same. This is a contradiction, because it violates the fact that the DD is reduced. Hence, the reduced, ordered DD in GF(N) must be unique.

Case 2 $\exists S \subseteq I_N$, such that $\forall k,l \in S$ and $k \neq l$, $u_k \neq u_l$. Therefore, by the induction hypothesis, $f^{u_k} \neq f^{u_l}$, $\forall k,l \in S$ and $k \neq l$. Let w be a node with $var(w) = x_1$, and $child_q(w) = u_q$, $\forall q \in I_N$.

Therefore,

$$f^w = \sum_{e=0}^{N-1} g_e(x_1) f^{u_e},$$

where $g_e(x_1) = 1 - [x_1 - \delta(e)]^{N-1}$, and the DD in GF($N$) rooted by w is reduced.

```
1     Algorithm ReduceDD(ν : node) : node;
2     begin
3        if ν is a terminal node, then
4            if IsIn(ν,HT), then return LookUp(HT,ν)
5            else begin
6                HT : = insert(ν,HT); (*Initially HT = ∅*)
7                return ν
8            end;
9        else begin
10           νᵢ' : = ReduceDD(childⱼ(ν)),∀i∈ Iₙ;
11           if νⱼ' = νₖ',∀j,k∈ Iₙ and j ≠ k, then return ν₀
12           else if IsIn(ν,HT), then return LookUp(HT,ν)
13           else begin
14               HT: = insert(ν,HT);
15               return ν
16           end
17       end
18    end;
```

Figure 7.25   A reduction algorithm for MODDs in GF(N)

By assumption $f^{u_t} = f|_{x_1} = \delta(t)$, $\forall t \in I_N$. Therefore, from Eq. 7.21, $f^w = f$.

Suppose that this reduced, ordered DD in GF(N) is not unique, and that there exists another reduced, ordered DD in GF(N) for f rooted by w'. Now, $f^{w'} = f$, and it must be the case that $f^{w'}|_{x_1} = \delta(k) \neq f^{w'}|_{x_1} = \delta(l)$, $\forall k,l \in S$, and $k \neq l$. If this is not the case, then $f|_{x_1} = \delta(i) = f^{w'}|_{x_1} = \delta(i) = f^{child_i}(w') = f^{child_j}(w') = f^{w'}|_{x_1} = \delta(j) = f|_{x_1} = \delta(j)$, $\forall i, j \in I_N$ and $i \neq j$. This is a contradiction, since this would imply that all the children of w' would be the same, thus violating the fact that the DD is reduced.

Now, owing to the ordering, $var(w') = x_1 = var(w)$. In addition, since $f^{w'} = f$, it follows that $f^{child_r}(w') = f|_{x_1} = \delta(r) = f^{u_r} = f^{child_r}(w)$, $\forall r \in I_N$. Therefore, by the induction hypothesis, $child_r(w') = u_r = child_r(w)$, $\forall r \in I_N$. Therefore, it follows that $w = w'$, by induction. Hence the proof for uniqueness follows by induction.

Minimality We now prove the minimality of a reduced, ordered DD GF(N) in terms of the total number of nodes, with respect to the reduction rules. Suppose that the reduced, ordered DD in GF(N) for f is not minimal with respect to the reduction rules. Then we can find a smaller DD in GF(N) for f as follows. If the DD in GF(N) contains a node v with $child_i(v) = child_j(v)$, $\forall i, j \in I_N$ and $i \neq j$, then eliminate v and for any node w with $child_k(w) = v$ ($k \in I_N$), make $child_k(w) = child_0(v)$.

If the DD in GF(N) contains distinct but isomorphic sub-graphs rooted by v and v', then eliminate v' and for any node w such that $child_k(w) = v'$ ($k \in I_N$), make $child_k(w) = v$.

A reduction algorithm

A reduction algorithm for MODD appears in Fig. 7.25. In order for sharing of equivalent sub-graphs, sub-graphs already present in the MODD are placed in a table. In lines 4, 6, 12, and 14 we have assumed that checking for membership and addition of an element to such a table can be carried out in constant time, e.g., by using a hash table (HT). Hence, the complexity of the algorithm is $O(|G|)$ since each node can be made to be visited just once during the reduction process, where G is an MODD of a function before the reduction.

However, an algorithm for the creation of MODDs from the functional description in GF(N) will have a complexity $O(N^n)$ in the worst case. The algorithm for the creation of MODDs can be derived from Eq. 7.20. The efficiency of the algorithm can be improved in general by noting that during the recursive expansion with respect to each variable, certain variables may not appear for further recursive calls. Therefore, the recursion tree need not be expanded in the direction of a variable that does not appear. The efficiency can be further improved by incorporating Lemma 7.20 based on a dynamic programming-like approach, as discussed in Section 7.5.6, which has been done for the experimental results in Section 7.5.11. In this case, the network in GF(N) is traversed in the topological order from the inputs to the outputs and Lemma 7.20 is applied iteratively.

Note that the size of a reduced MODD depends heavily on the variable ordering, as in any other DD [1,81]. The depth of an MODD is $O(n)$ in the worst case since each variable appears once at each level in the worst case.

7.5.5   Variable reordering

The size, i.e., number of nodes, of an MODD depends on the order of the variables during its construction. For example, the MODD in Figure 7.24(a) represents a function in GF(3) under the variable order (x_1, x_2). Figure. 7.24(b) shows the same function in GF (3), but under the variable order (x_2, x_1). Clearly, Fig. 7.24(a) contains fewer nodes than Fig. 7.24(b). Given an n variable function in GF(N), the size of the solution space for finding the best variable order is $O(n!)$, which is impractical for large values of n. Hence, a heuristic level-by-level swap-based algorithm is considered in this chapter.

The theory behind variable reordering in GF(N) is based on Theorem 7.12 as follows. Without loss of generality, let us assume that variables x_1 and x_2 are to be swapped. From Theorem 7.12 we have,

$$f(x_1, x_2 \ldots, x_n) = g_0(x_1)(g_0(x_2)f|_{0,0} + g_1(x_2)f|_{0,1} + \cdots + g_{N-1}(x_2)f|_{0,N-1})$$
$$+ g_1(x_1)(g_0(x_2)f|_{1,0} + g_1(x_2)f|_{1,1} + \cdots + g_{N-1}(x_2)f|_{1,N-1}) + \cdots$$
$$+ g_{N-1}(x_1)(g_0(x_2)f|_{N-1,0} + g_1(x_2)f|_{N-1,1} + \cdots + g_{N-1}(x_2)f|_{N-1,N-1}).$$

If x_1 and x_2 are swapped, then, again from Theorem 7.12, the following function f_s results:

$$f(x_1, x_2 \ldots, x_n) = g_0(x_2)(g_0(x_1)f|_{0,0} + g_1(x_1)f|_{1,0} + \cdots + g_{N-1}(x_1)f|_{N-1,0})$$
$$+ g_1(x_2)(g_0(x_1)f|_{1,0} + g_1(x_1)f|_{1,1} + \cdots + g_{N-1}(x_1)f|_{N-1,1}) + \cdots$$
$$+ g_{N-1}(x_2)(g_0(x_1)f|_{0,N-1} + g_1(x_1)f|_{1,N-1} + \cdots + g_{N-1}(x_1)f|_{N-1,N-1}),$$

where $f \equiv f_s$. It can be noted by comparing f and f_s that in f_s each literal $g_e(x_1)$ $(g_e(x_2))$ is swapped with $g_e(x_2)$ $(g_e(x_1))$ for $e \in I_N$, and each co-factor $f|_{r,s}$ $(f|_{s,r})$ is swapped with $f|_{s,r}$ $(f|_{r,s})$. Figure 7.26 shows the MODDs corresponding to f (top MODD) and f_s (bottom MODD). In the top MODD, variables x_1 and x_2 appear in levels 0 and 1 respectively, while the co-factors appear as external nodes. This can be a more general case, e.g., the two variables may appear in any arbitrary but consecutive levels in a larger MODD, e.g., variables x_i and x_{i+1} and $2 < i \le n$. Clearly, to swap two variables all we have to do is: (a) swap the contents of the nodes (i.e. the variables) in the two levels, and (b) swap each

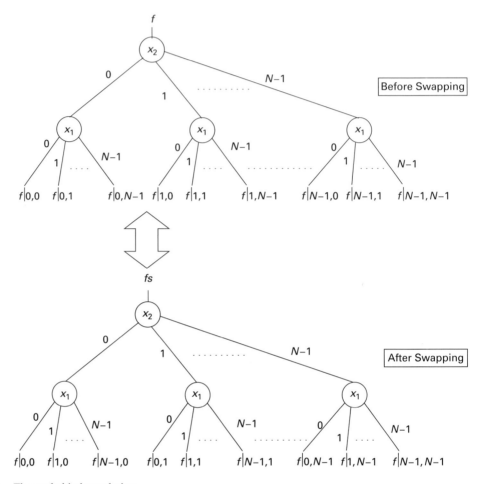

Figure 7.26 Theory behind reordering

co-factor $f|_{r,s}$ ($f|_{s,r}$) with $f|_{s,r}$ ($f|_{r,s}$). However, care must be exercised when a level has one or more missing nodes from reduction. If this happens, the missing nodes may have to be recreated in the swapped version. Also, some nodes may become redundant after the swap, in which case the redundant nodes must not appear in the final result (refer to Example 7.4). However, if a node at level$-i$ ($0 \leq i < n$), which is to be swapped with level$-i + 1$, does not have any children at level$-i + 1$, then it can be moved to level$-i + 1$ directly. By similar reasoning, if a node at level$-i + 1$ does not have any parent nodes at level$-i$, then that node can be moved up to level$-i$ directly. Note that swapping two levels i and $i + 1$ does not affect the other levels, i.e., those in the range $0 \leq j < i$ (if $i > 1$) and $i + 1 < k < n$ (if $i < n-1$).

The heuristic swap-based variable reordering algorithm presented in the following is based on this (Theorem 7.12). A swap-based variable reordering algorithm exists for BDD, [8] but it is not suitable for MODD reordering.

The algorithm uses an array of hash tables, where the array indexes correspond to the levels in an MODD for direct access to each of the nodes within a level. It proceeds by sifting a selected level (i.e., a variable) up or down by swapping it with a previous or next level. The level with the largest number of nodes is considered first, and then the one with the next largest node count, and so on, i.e., the array of hash tables is sorted in descending order of the hash-table sizes. Once the level to be sifted first is considered it is sifted up if it is closer to the root, or down if it is closer to the external nodes. If it lies in the middle, then the decision to sift either up or down is made arbitrarily. The algorithm stops after a complete sift-up-and-down operation. The complexity of the algorithm can be argued to be $O(n^2)$. [8] Various heuristics have been considered to limit the sift and swap operations for speed up the algorithm. For example, if a sift-up (down) operation doubles the node count, no more sift-up (down) operations are carried out.

Example 7.4 Figure 7.27 shows the basic idea behind the swap algorithm. Fig. 7.27(a) shows the original MODD for a function $f(x_1,x_2,x_3)$ in GF(4) generated by recursive expansion of Theorem 7.12. Here, variables x_1,x_2 and x_3 appear in levels 0, 1, and 2 respectively. Level 1 (i.e., variable x_2) contains the largest number of nodes. Hence, this is considered to be the starting point of the sift operation. Level 1 is equidistant from level 0 and level 2. Hence, a sift-up is chosen arbitrarily. Swapping between level 0 and 1 results in Fig. 7.27(b). The nodes shown with broken lines are redundant owing to the fact that all their children point to the same node (Corollary 7.17). Hence, these nodes are not considered in the final result of Fig. 7.27(c). For example, considering the paths with the edge $x_2 = 1$ in Fig. 7.27(b), all the paths with edge $x_2 = 1$ leading to node x_3 in Fig. 7.27(a) have the edges with variable $x_1 = i$ for $i = 0,1,a,\beta$. Therefore, node x_1 becomes redundant by Corollary 7.17, if node x_1 appears after node x_2 under this circumstance. The same reasoning applies to the other two nodes with variable x_1 shown with the broken lines.

It can be shown that further sift operations do not yield additional node reduction. The original node count was five, and the new node count is three.

7.5.6   Operations in GF(*N*)

Algebraic operations

Algebraic operations, such as addition, multiplication, subtraction, and division, in GF(*N*) can be carried out between two MODDs. Consider the following lemma, which can be shown to hold by perfect induction.

LEMMA 7.20 *Let $f(x_1, \ldots, x_i, \ldots, x_n)$ and $h(x_1, \ldots, x_i, \ldots, x_n)$ be two functions in* GF(*N*), *and let $\odot$ represent an algebraic operation in* GF(*N*). *Then*

$$f \odot h = \sum_{e=0}^{N-1} g_e(x_i)(f|_{x_i=\delta(e)} \odot h|_{x_i=\delta(e)}), \qquad (7.25)$$

where $g_e(x_i) = 1-[x_i-\delta(e)]^{N-1}$.

Lemma 7.20 can be implemented recursively to perform algebraic operations between MODDs. However, application of Lemma 7.20 directly will almost certainly be explosive

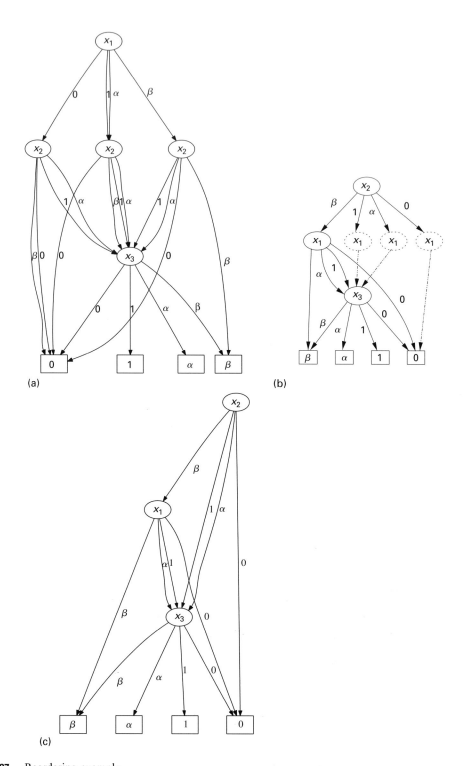

Figure 7.27   Reordering example

in terms of the search space. Two things can be done to eliminate this. Firstly, while the resulting DD in GF(N) is being constructed, it can be reduced at the same time. Secondly, intermediate results can be stored in a cache (dynamic programming), thus eliminating many operations, which will otherwise have to be repeated.

Let G_f and G_h be the reduced MODDs for f and h respectively. The complexity of such an operation can be reasoned about by considering the case for BDDs. [1] Assuming that insertion and deletion from the cache can be carried out in constant times, owing to the dynamic programming nature the number of recursive calls can be limited to $O(|G_f| \cdot |G_h|)$.

Composition
We have the following, which can be shown to hold by perfect induction.

LEMMA 7.21 *Let $f(x_1, x_2, \ldots, x_i, \ldots, x_n)$ and $h(x_1, x_2, \ldots, x_n)$ be two functions in* GF(N). *Then,*

$$f|_{x_i=h} = \sum_{e=0}^{N-1} \left[1 - (h - \delta(e))^{N-1} \right] f|x_i = \delta(e). \tag{7.26}$$

An algorithm for composition of two functions in GF(N) can be formed based on Lemma 7.21 in a manner similar to that for a BDD. [1] For this operation, we require a *restrict operation* in GF(N), similar to that for a BDD, and the algebraic operations presented previously. The restrict algorithm can be constructed for a reduced ordered DD in GF(N) in a manner similar to that for a BDD, and is not shown here, for brevity.

Multiple-valued SAT
Given a function $f(x_1, x_2, \ldots, x_n)$ in GF(N) and $T \subseteq I_N - \{0\}$, the idea is to find an assignment for x_i, $\forall i \in \{1, 2, \ldots, n\}$, such that the value of f is $\{\delta(s) | s \in T\}$. If such an assignment exists, then f is said to be satisfiable (MV-SAT); otherwise it is unsatisfiable.

The MV-SAT problem finds applications in bounded model checking, simulation, testing, and verification. An algorithm for *any* such satisfying assignment will have a complexity $O(|G_f|)$, where G_f is the reduced MODD for the function f in GF(N). An algorithm for *all* such assignments would have an exponential complexity. However, this process can be speeded up by considering characteristic and encoded characteristic functions in GF(N) and their evaluation times as discussed in Sections 7.5.9 and 7.5.10. [92,93]

7.5.7   Multiple-output functions in GF(N)

For multiple-input multiple-output binary functions, the inputs or outputs can be arbitrarily grouped into m-bit chunks and each m-bit chunk can be represented in GF(2^m) with a single MODD. Further node reduction can be obtained by sharing the nodes between each of the MODDs representing a chunk of bits. Such an MODD is called a *shared MODD* or SMODD. The general idea is shown in Fig. 7.28. The SMODD is, basically, a single diagram with multiple root nodes, which is also canonic. The canonicity of the SMODD can be argued in a similar manner as for a single MODD.

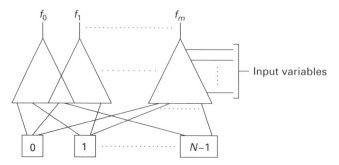

Figure 7.28   General structure of shared MODD

Similar reasoning can be carried over to higher-order fields. Given any multiple-output function in GF(R), where R is a power of a prime, the inputs and outputs can be arbitrarily grouped into m R-valued chunks and each chunk can be represented in GF(R^m) by means of an MODD. Then an SMODD will represent all the chunks simultaneously.

The concept of levels is applicable to SMODDs across all the outputs simultaneously by trivial reasoning. Therefore, the theory behind variable reordering, as discussed in Section 7.5.5, applies equally well to SMODDs. In this case, when levels i and $i + 1$ ($0 \leq i < n$) are swapped, all the nodes in levels i and $i + 1$ across all the outputs have to be considered simultaneously. Therefore, the swap-based sift-reordering algorithm discussed in Section 7.5.5 works equally well for SMODDs and MODDs.

7.5.8   Further node reduction

Further node reduction can be obtained by means of the following two rules, in addition to the two rules presented in Section 7.5.3.

- *Zero suppression* Suppress the 0-valued terminal node, along with all the edges pointing to it.
- *Normalization* Move the values of the non-zero terminal nodes as weights to the edges, and ensure that (a) the weight of a specific valued edge (e.g., that with the highest value) is always 1, and (b) assuming P represents the set of all the paths, $\forall z \in P$ the GF(N) product of all the weights along z is equal to the value of the function corresponding to z.

Note that the zero suppression rule is unlike the reduction rule for the zero-suppressed BDD. [96] It can be argued that the above two rules will also maintain the canonicity if the weights are assigned in a fixed order throughout the graph during normalization. A reduced graph obtained using the above four reduction rules in GF(N) will be called a *zero-suppressed normalized MODD* or a ZNMODD. The values of the terminal nodes in an MODD are *distributed* as weights over each path in the ZNMODD. To read a value of a function from a ZNMODD, first the path corresponding to the inputs is determined. Then all the weights along that path are multiplied in GF(N), which

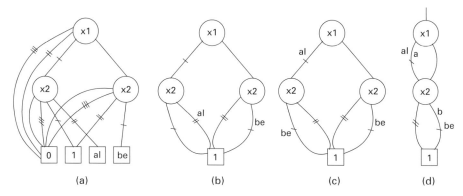

Figure 7.29   Example of ZNMODD reduction

corresponds to the value of that function. In the rest of the chapter, the weight of the highest-valued edge will be normalized to 1, unless otherwise stated.

Example 7.5 Let us consider the function $f(x_1,x_2) = [0\beta1001a000000000]$ in GF(4), where $\{0,1,a,\beta\}$ are the elements of GF(4). Figure 7.29(a) shows this function realized by means of a reduced MODD. Figs. 7.29(b)–(d) show the gradual conversion to ZNMODD. Here, the lines with zero, one, two, and three cuts represent the values 0, 1, a, and β respectively. Note how the weights are moved around and adjusted.

In Fig. 7.29(b) the terminal node with 0-value is suppressed along with all the edges pointing to it. Also the non-zero values of the terminal nodes are moved as weights associated with the terminal edges. Let us normalize with respect to the highest-valued edge, i.e., make the weight of the highest valued edges, β in this case, equal to 1. The a-edge of the left sub-graph rooted at x_2 has a weight of a. Therefore, to make its weight equal to 1, a is moved up one level, while the 1-edge is assigned a weight of β to maintain the correctness of the underlying function. This results in the ZNMODD of Fig. 7.29(c). Clearly, in Fig. 7.29(c) the two sub-graphs rooted at x_2 are isomorphic, which can be shared resulting in the ZNMODD of Fig. 7.29(d).

Now, let us find the value of $f(1,1)$. This should yield a 1. From Fig. 7.29(d), this corresponds to the path ab. The value of this function is, therefore, $1 \cdot a \cdot \beta = a \cdot \beta = 1$. Similarly, $f(1,a)$ yields $1 \cdot a \cdot 1 = a$, and so on.

Note that the total number of paths in Fig. 7.29(a) is ten, while that in Fig. 7.29(d) is only four.

7.5.9   Representing characteristic functions in GF(N)

The characteristic function (CF) defines a relation over inputs and outputs, such that CF $= 1$ if for a specific input combination the output is valid; otherwise CF $= 0$.

Let us consider a multiple-output function defined over finite fields: $f(x_1,x_2,\ldots,x_n) = (y_1,y_2,\ldots,y_m)$. Let $X = (x_1,x_2,\ldots,x_n)$ and $Y = (y_1,y_2,\ldots,y_m)$. Then the $(n + m)$-input 1-output CF is defined as

$$\phi\left(X,Y\right) = \begin{cases} 1 & \text{if } f(X) = Y \\ 0 & \text{otherwise.} \end{cases}$$

An SMODD can be constructed from the above, which will constitute the n input variables and m *auxiliary* variables (AV) corresponding to each of the outputs. Such an SMODD will be called a CF-SMODD. Given an input combination of f, the nodes corresponding to the AVs in the CF-SMODD decide the outputs of f. For each node corresponding to an AV, except for only one edge, all the edges lead to the terminal node 0. The edge leading to the non-zero terminal node determines the output of the function. Examples of the CF can be found in [97].

The concept of CF can be extended by allowing output encoding, since there is only one possible set of outputs for a given input combination. The resulting function can be represented by a mapping $\eta:G^n \times G^l \to G$, where $l = \lfloor \log_N(m) \rfloor$ and will be called the *encoded* CF or ECF. The ECF has l AVs. Each output is defined by one of the N^l input combinations in an l AV function. As with the CF-SMODD, an ECF can be represented by means of an SMODD, which we shall call the ECF-SMODD. The following example illustrates the key points.

7.5.10   Evaluation of functions

Example 7.6 Let us consider a five-input three-output binary function defined as follows, with the inputs denoted by the variables (x_0,x_1,x_2,x_3,x_4) and the outputs denoted by (f_0,f_1,f_2).

$$f_0 = \sum m(15,23,26,29,30,31)$$
$$f_1 = \sum m(1,2,4,8,11,13,14,16,21,22,26,31)$$
$$f_2 = \sum m(3,5,6,7,8,9,14,11,12,13,17,19,20,21,22,23,24,25,26,28).$$

Let us assume that the function is encoded in GF(4) with inputs and outputs grouped as $X_0 = (x_0,x_1)$, $X_1 = (x_2,x_3)$, $X_2 = x_4$, $F_0 = (f_0,f_1)$, $F_1 = f_2$. The CF of this function is $\varphi(X_0,X_1,X_2,F_0,F_1)$. We have assumed that the binary combinations $10 = a$ and $11 = \beta$. The variables F_0 and F_1 can be encoded as $A_0 = 0$ for F_0 and $A_0 = a$ for F_1, resulting in the ECF $\eta(X_0,X_1,X_2,A_0)$. Note that we can encode four functions using one AV. Here we have only two functions. The resulting ECF-SMODD appears in Fig. 7.30(a).

The evaluation of the SMODDs is required to find a satisfying assignment corresponding to an input pattern. The path corresponding to the given input pattern is traced from the root node to one of the terminal nodes, and the value of the terminal node gives the satisfying assignment, if it exists. This is an $O(n)$ operation, which can become a bottleneck, especially when the number of inputs is large and there are many input patterns to be evaluated. However, fast evaluation is highly desirable for applications in simulation, testing, and safety checking [92,93].

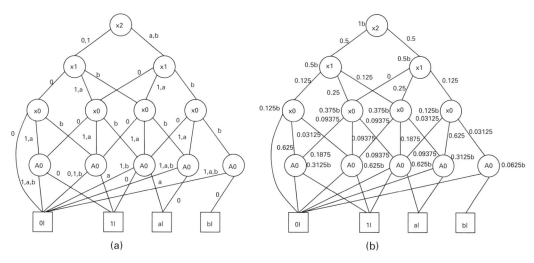

Figure 7.30   (a) ECF-SMODD, (b) computation of APL

In the case of a CF, the outputs are evaluated at the AVs. As we know, all the outgoing edges except only one edge lead to the zero terminal node. The remaining edge indicates the value of the function. With the ECF, once we reach the node corresponding to the AV, the paths corresponding to each of the output encodings are traced to find the value.

For example, let us consider an input pattern $(x_0 = 1, x_1 = 1, x_2 = 0, x_3 = 0, x_4 = 1)$ for the ECF-SMODD of Fig. 7.30. The pattern is $(X_2 = a, X_1 = 0, X_0 = \beta)$ if it is encoded in GF(4) (Example 7.6). The path traced by the evaluation is shown in bold. After reaching node A_0 in the path, the path corresponding to the given encoding for each output has to be taken into account. In this case, the encoding was defined as $A_0 = 0$ for F_0 and $A_0 = a$ for F_1. Hence, if we take the path corresponding to $A_0 = 0$ we end up at terminal node 1, thus giving us $F_0 = 1$. Similarly, $F_1 = a$. This results in $(f_0 = 0, f_1 = 1, f_2 = 1)$, which is the required evaluation.

Comparison of evaluation times
Functions can be represented and evaluated by means of SMODD, CF-SMODD, or ECF-SMODD. This section provides a mechanism for comparing the evaluation times for each of the cases. We have tested our theory on many benchmarks, and the results appear in Section 7.5.11. A good estimation of evaluation times can be obtained by computing the *average path length* (APL), which we define below.

1. *Node-traversing probability ($P(V_i)$)* The probability of traversing the node V_i when an MODD is traversed from the root node to a terminal node.
2. *Edge-traversing probability ($P(e_{j,v_i})$)* The probability of traversing the edge $j = 0, 1, \ldots, p^{m-1}$ from the node V_i, i.e. $P(e_{j,v_i}) = \frac{P(V_i)}{p^m}$, for nodes corresponding to the input variables.
3. The edge-traversing probability for edges emanating from a node corresponding to an AV is $P(e_{j,v_i}) = P(V_i)$ since all the edges have to be traversed for determining all the outputs of the function.

Algorithm ENC_PROB(int *NODE*, float *PROB*)
{   // *NODE* is the current node, and *PROB* is its probability
    if (*NODE* = Terminal Node) return 0;
    Add *PROB* to *NODE*'s probability; // Each node stores its probability
    Decrease the Reference count of *NODE* by 1;
    if (Reference count of *NODE* > 0) return 0;
    float *TOT* = 0;
    for(int $i = 0; i < N_o; i++$){ // N_o is the output field size
        T = child$_i$(*NODE*);
        if(*NODE* = auxiliary node){
            clear reference counts for all nodes of the subtree T;
            make new reference count only for T;
            clear probability variable in all nodes of T;
            APL for T = ENC_PROB(T, *NODE*'s probability);
            TOT = *TOT* + APL for T + *NODE*'s probability;
        }else{
            APL for T = ENC_PROB(T, *NODE*'s probability/N_o);
            TOT = *TOT* + APL for T;
        }
    }
    if(*NODE*! = auxiliary node) *TOT* = *TOT* + *NODE*'s probability;
    return *TOT*;
}

Figure 7.31   Algorithm for calculation of APL in ECF-SMODD

4. The node-traversing probability is equal to the sum of all the edge-traversing probabilities incident on it.
5. *Average path length (APL)* For an SMODD, the APL is equal to the sum of the node-traversing probabilities of the non-terminal nodes. For a CF-SMODD and ECF-SMODD, the APL is equal to the sum of the node traversing probabilities of those nodes above the AVs, and the APLs for each sub-graph rooted at the AVs.
6. The average path length of a shared MODD is the sum of the average path lengths of the individual MODDs.

An algorithm for computing the APL for ECF-SMODD appears in Fig. 7.31. Algorithms for computing the APLs for SMODD and CF-SMODD can be formulated from this algorithm, which we have implemented in Section 7.5.11, but the details have been left out for brevity.

For example the node- and edge-traversing probabilities of the ECF-SMODD in Fig. 7.30(a) appear in Fig. 7.30(b). The first three levels in the tree correspond to the input variables. Hence, the probabilities are computed using definitions 1, 2, and 4. However, since all the outputs have to be considered, the APL for each sub-tree is separately computed at the auxiliary nodes. The probabilities at the auxiliary nodes correspond to the sums of the APLs of each sub-tree of the outputs. The APL is $1 + (0.5 + 0.5) + (0.125 + 0.375 + 0.375 + 0.125) + (0.3125 + 0.625 + 0.625 + 0.3125 + 0.0625) = 4.9375$.

7.5.11   Experimental results

The techniques in this chapter have been applied to a number of benchmarks, including integer-multiplier circuits. The program was developed in C++ (Gnu C++ 3.2.2) and tested on a Pentium-4 machine with 256MB RAM, running RedHat Linux 9 (kernel 2.4).

Performance

Table 7.6 shows results from the standard IWLS'93 and MCNC benchmark sets. Columns "I/P" and "O/P" represent the total number of inputs and outputs, column "SBDD" shows the number of nodes obtained using the shared ROBDD representation, while column GF(2^r) represents the number of nodes obtained based on the proposed DD for a field size of 2^r. In column GF(2^r), r adjacent bits are grouped together for each variable in GF(2^r). The nodes of the proposed DD are also shared across the outputs. The same notation is used for the other tables.

The columns with the headings "W/o reord" and "Reord" present the node count without and with variable reordering. A first-come, first-served variable ordering is considered for the "W/o reord" columns. For the reordering, the variable-by-variable swap-based sifting algorithm of Section 7.5.5 has been employed. Significant node reduction is apparent for many circuits, e.g., misex3c. However, in some cases, the node count has increased, owing to the lack of sharing. Note that successive swap operation in a particular direction (up or down) is only carried out if a swap operation does not increase the size of the MODD by two or more. This restriction can be relaxed (e.g., by considering the maximum allowable size increase to be 1.5 times) to obtain better results sometimes. However, this also increases the execution time, as more swaps are carried out. Variable reordering algorithms exist for MDDs. [98] However, they cannot be directly compared with the presented technique because the presented technique has been applied to varying input and output field sizes, while [98] seems to have ignored this aspect.

Apart from the benchmark 9sym, all the circuits have been tested up to GF(16). Benchmark 9sym is a single-output circuit and, hence, its testing in higher-order fields seemed to be unjustified. In Table 7.6, the input and output field sizes are kept the same. Clearly, as we move to a higher-order field the number of nodes is reduced for the majority of the cases. Also, as we move to a higher-order field, node reduction owing to reordering seems to be more effective.

For the rest of the tables, apart from Table 7.9, reordering has not been done, to illustrate the other properties of the MODD more effectively, e.g., the effect on the node count when the input and output field sizes are varied, and when the MODDs are ordered based on the evaluation times.

Table 7.7 represents the results for the same set of benchmarks under the same variable ordering but the input field size is varied while the output field size is kept at constant 2. Clearly, the number of nodes has reduced further for many benchmarks as compared with Table 7.6. The reason seems to be improved sharing of nodes between the different outputs.

Table 7.6 Same field size for input and output

Benchmark	I/P	O/P	SBDD	GF(2^2)		GF(2^3)		GF(2^4)	
				W/o reord	Reord	W/o reord	Reord	W/o reord	Reord
5×pl	7	10	88	48	42	43	35	27	16
9sym	9	1	33	17	17	10	10	–	–
apex 4	9	19	1021	536	515	324	324	241	136
bl2	15	9	91	78	47	48	45	61	51
bw	5	28	118	76	72	59	47	59	21
clip	9	5	254	136	89	88	41	49	31
misex 3c	14	14	844	505	279	396	181	586	156
duke2	22	29	366	793	507	783	445	601	453
table5	17	15	685	751	678	636	636	499	348
e64	65	65	194	943	569	858	601	624	495
cordic	23	2	75	29	28	21	20	15	15
misex 2	25	18	100	93	81	50	42	45	41
pdc	16	40	596	433	310	384	384	212	212
spla	16	46	628	352	339	261	261	155	155
exlOlO	10	10	1402	662	654	346	344	670	207

Table 7.7 I/P field varying with constant output field size of 2

Benchmark	I/P	O/P	SBDD	$GF(2^2)$	$GF(2^3)$	$GF(2^4)$
5xpl	7	10	88	57	51	43
9sym	9	1	33	18	16	22
apex4	9	19	1021	509	346	240
bl2	15	9	91	69	56	60
bw	5	28	118	71	47	42
clip	9	5	254	149	106	69
misex3c	14	14	844	450	300	246

Table 7.8 O/P field size varying with constant input field size of 2

Benchmark	I/P	O/P	SBDD	$GF(2^2)$	$GF(2^3)$	$GF(2^4)$
5xpl	7	10	88	87	77	87
9sym	9	1	33	–	–	–
apex4	9	19	1021	1031	989	983
bl2	15	9	91	102	84	126
bw	5	28	118	154	148	140
clip	9	5	254	227	211	195
misex3c	14	14	844	903	978	1333

Table 7.9 Same field size for both input and output

Multiplier	SBDD	$GF(2^2)$		$GF(2^3)$		$GF(2^4)$	
		W/o reord	Reord	W/o reord	Reord	W/o reord	Reord
2*2	14	7	7	6	5	1	–
3*3	51	28	28	15	15	16	9
4*4	157	98	87	84	60	31	31
5*5	471	254	249	183	183	272	121
6*6	1348	795	731	736	624	431	428

Table 7.8 shows the result with input field size kept at constant 2 and the output field size varied, again under the same variable ordering. In general, the number of nodes seems to have increased, owing to lack of sharing between the different outputs. In other words, higher output field size seems to hamper sharing of nodes between different outputs, even though the number of nodes in each output may reduce. This observation seems to be consistent with the conclusion drawn from Table 7.7.

Table 7.9 shows results for $n*n$ integer multipliers for $n = 2,3,4,5,6$. The input and output field sizes are kept the same in this table. Clearly a substantial reduction in the

Table 7.10 I/P field varying with constant output field size of 2

Multiplier	SBDD	GF(2^2)	GF(2^3)	GF(2^4)
2*2	14	9	10	4
3*3	51	33	24	28
4*4	157	78	64	55
5*5	471	263	190	127
6*6	1348	695	389	465

Table 7.11 O/P field size varying with constant input field size of 2

Multiplier	SBDD	GF(2^2)	GF(2^3)	GF(2^4)
2*2	14	15	11	12
3*3	51	50	65	46
4*4	157	178	194	260
5*5	471	490	553	755
6*6	1348	1587	1917	1890

number of nodes is noticeable for the majority of the benchmarks. In some cases, reordering has produced further improvement, e.g., the 4 * 4 and 6 * 6 multipliers in GF (8). Although we have not explicitly shown the results in GF(2^5) and GF(2^6), MODD reported only 63 nodes as opposed to the 471 nodes for SBDD for the 5 * 5 multiplier in GF(2^5). Also, for the 6 * 6 multiplier, the number of nodes reported by the MODD in GF(2^6) is only 127 as opposed to 1348 for the SBDD, i.e., more than an order of magnitude reduction. Also, this table suggests that the node reduction seems to improve as we consider larger and more practical integer multipliers.

Table 7.10 shows the results with varying input field size and a constant output field size of 2. Again, considerable improvement has been observed for some benchmarks, owing to the improved sharing of the nodes across the outputs.

Table 7.11 shows the results for the multipliers with fixed input field size and varying output field size. As anticipated, the number of nodes has increased owing to the possible lack of sharing.

As we move to a higher-order field from a lower-order field, the number of nodes usually decreases. This decrease is also associated with a smaller number of levels and shorter path lengths than conventional BDDs and their variants.

Evaluation time

Table 7.12 shows the results for the APLs compared with those for SBDDs. For the majority of cases, the APLs are significantly lower than those in the SBDDs. Also, as we go from GF(4) to GF(8), the APLs reduce further. On average, the APLs are three times less in GF(4) and about six times less in GF(8) as compared with the SBDDs. That is, the evaluation time essentially halves as we go from GF(4) to GF(8). Note that

Table 7.12 APL compared with SBDD

| | | | BDD | | SMODD | | | | | |
| | | | | | GF(4) | | | GF(8) | | |
Benchmark	I/P	O/P	Nodes	APL	Nodes	APL	% Imp.	Nodes	APL	% Imp.
duke2	22	29	366	150.3	793	80.61	186	783	48.87	307
cordic	23	2	–	–	29	4.41	–	21	3.29	–
misex2	25	18	100	75.6	93	16.55	456	50	8.61	877
vg2	25	8	90	48.9	892	24.15	202	867	17.2	284
table5	17	15	685	114.1	751	32.12	355	635	16.02	712
pdc	16	40	596	215.4	433	52.58	409	384	33.13	650
e64	65	65	194	256	943	64.96	394	858	56.27	454
spla	16	46	628	226.6	352	47.33	478	261	30.18	750
exIOIO	10	10	–	–	662	24.69	–	346	14.74	–
ex5	8	63	–	–	194	66.72	–	144	36.05	–
bl2	15	9	–	–	78	15.34	–	48	8.31	–
x6dn	39	5	235	41.2	212	7.26	567	165	4.09	100
exep	30	63	675	255.7	462	53.86	474	409	33.44	764
xldn	27	6	139	41	151	10.3	397	120	6.56	624
x9dn	27	7	139	50.4	160	12.6	399	174	10.66	472
markl	20	31	119	115.7	149	36.95	313	150	26.63	434
mainpla	27	54	1857	277.5	2414	138.26	200	1667	70.74	392
risc	8	31	–	–	58	23.72	–	32	13.33	–
xparc	41	73	1947	304.8	1925	94.46	322	1550	58.22	523
apex4	9	19	–	–	536	36.54	–	324	16.73	–
Average	23.5	29.7	555	155.23	564.35	42.17	368	449.45	25.65	524.5

Table 7.13 APL with random pattern simulation in GF(4)

Benchmark (IP/OP)	SMODD			CF-SMODD			ECF-SMODD		
	Nodes	APL	Random simulation	Nodes	APL	Random simulation	Nodes	APL	Random simulation
duke2 (22/29)	793	80.61	4023605	445	8.1	405126.5	470	28.03	439206.5
cordic (23/2)	29	4.41	220771.5	30	5.14	256775	28	4.17	83466.8
misex2 (25/18)	93	16.55	825155	131	6.65	317410	83	4.22	84350.6
table5 (17/15)	751	32.12	1605105	555	6.51	326386.5	604	14.67	201920
pdc (16/40)	433	52.58	2630225	1356	9.28	464182	418	63.54	1200000
e64 (65/65)	943	64.96	3251360	66	1.33	66643.5	442	13.24	226700
spla (16/46)	352	47.33	2384420	1454	7.26	356000	345	47.93	961395
exlOIO (10/10)	662	24.69	1234420	555	7.29	343225	655	36.55	600000
ex5 (8/63)	192	66.72	3337715	980	20.77	1083580	205	110.72	2200000
bl2 (15/9)	78	15.34	767230	157	7.11	265646.5	62	21.6	400000
x6dn (39/5)	212	7.26	363796.5	201	5.4	265861.5	144	3.08	5012.8
exep (30/63)	462	53.86	2695070	1459	10.05	493531.5	423	23.75	509220
xldn (27/6)	151	10.3	515495	358	6.73	332395	127	5.92	2.71
x9dn (27/7)	160	12.6	628695	456	7.95	397483	129	6.47	129333.4
markl (20/31)	149	36.95	1850300	281	8.69	426828.5	116	18.4	323713.5
mainpla (27/54)	2414	138.26	6914400	2153	9.56	468033.5	811	41.51	1083405
risc (8/31)	58	23.72	1185410	64	4.97	248535	55	14.91	215953.5
xparc (41/73)	1925	94.45	4732915	2537	5.31	260840	983	16.63	337326.5
apex4 (9/19)	536	36.54	1825970	1078	10.81	540455	505	17.7	204093.5

Table 7.14 APL with random pattern simulation in GF(8)

Benchmark (IP/OP)	SMODD			CF-SMODD			ECF-SMODD		
	Nodes	APL	Random simulation	Nodes	APL	Random simulation	Nodes	APL	Random simulation
duke2 (22/29)	783	48.87	2444800	373	6.53	326456.5	412	29.15	637575
cordic (23/2)	21	3.29	165215	22	3.68	184286.5	20	2.7	90145
misex2 (25/18)	50	8.61	429593.5	88	4.2	204385	43	16.63	400000
table5 (17/15)	635	16.02	801950	379	4.48	224104.5	247	4.28	207866.5
pdc (16/40)	384	33.13	1656360	1096	5.62	281190	387	57.13	1200000
e64 (65/65)	858	56.27	2813180	44	1.14	57121.5	443	3.01	75326.25
spla (16/46)	261	30.18	1509950	1132	3.77	188649	259	31.6	848145
exlOIO (10/10)	346	14.74	737095	337	5.8	274060	345	22.73	400000
ex5 (8/63)	144	36.05	1802605	652	13.28	663865	148	68.05	1600000
bl2(15/9)	48	8.31	415075	120	5.43	265646.5	46	16.05	400000
x6dn (39/5)	165	4.09	206140	232	2.71	134725	151	2.45	81764.33
exep (30/63)	409	33.44	1674265	1073	6.54	327246.5	368	157.18	4536146.5
xldn (27/6)	120	6.56	327983.5	302	5.23	259120	121	14.56	400000
x9dn (27/7)	174	10.66	532785	216	5.27	263295.5	129	5.16	129103.75
markl (20/31)	150	26.63	1850300	196	6.53	332668.5	116	139.87	800400
mainpla (27/54)	1667	70.74	3545165	1593	6.69	329915	537	91.71	2764640
risc (8/31)	32	13.33	666330	44	3.1	154980.5	35	27.16	800000
xparc (41/73)	1550	58.22	2914670	1820	3.8	188025	731	3.22	59733.25
apex4 (9/19)	324	16.73	837530	722	7.46	373047	325	24.73	400000

the number of nodes in the SMODDs is almost identical on average to that in the SBDDs. This is because the SMODDs have been ordered based on the APLs, which does not necessarily guarantee reduced node count. The dashes ('−') in the table indicate that results for those circuits (i.e., ex1010, ex5, b12, risc, apex4) are not available for the BDDs.

Tables 7.13 and 7.14 present a comparison of the APLs for SMODD, CF-SMODD, and ECF-SMODD for the benchmark circuits modeled in GF(4) and GF(8), respectively. These tables also show the results for 50 000 random vectors. The results for random pattern simulation constitute the net total path lengths for the 50 000 vectors. The spatial complexity is reflected by the node count and the speed of evaluation by the APLs and random pattern simulations. These results reflect a speed-up over current methods for simulation such as in [94]. The trade-off between these two factors across the representations is clearly evident from the results shown in these tables. In general, it can be seen that the CF-SMODD clearly wins out in terms of speed, whereas the ECF-SMODD tries to optimize between the speed and node count.

7.5.12   Conclusions

This chapter focused on a framework for representing multiple-output binary and word-level circuits based on canonic DDs in GF(N). We showed that such reduced ordered DDs are canonical and minimal with respect to a fixed variable ordering. Techniques for further node and path optimization have also been presented. We also presented the theory for representing functions in GF(N) in terms of their CF and ECF under the same framework.

The proposed DDs have been tested on many benchmarks with varying input and output field sizes. The results suggest superior performance in terms of node compression as well as reduced APLs, which implies improved evaluation times. This has also been confirmed by a simulation of 50 000 randomly selected vectors. Overall, the results seem to suggest that the proposed framework can serve as an effective medium for verification as well as for simulation, testing, and safety checking.

7.6 Acknowledgements

The work on TEDs described in this chapter has been supported by a grant from the National Science Foundation under award No. CCR-0204146, INT-0233206, and CCR-0702506. The work on FFDDs has been funded, in part, by the Engineering and Physical Science Research Council, UK Grant No. GR/S40855/01.

7.7 References

[1] R. E. Bryant (1986). Graph-based algorithms for Boolean function manipulation. *IEEE Transactions Computers*, **C–35**(8):691–677.

[2] R. E. Bryant and Y. A. Chen (1995). Verification of arithmetic functions with binary moment diagrams. In *DAC–95*.

[3] M. Ciesielski, P. Kalla, and S. Askar (2006). Taylor expansion diagrams: a canonical representation for verification of dataflow designs. *IEEE Transactions on Computers*, **55**(9):1188–1201.

[4] A. M. Jabir, D. K. Pradhan, A. K. Singh, and T. L. Rajaprabhu (2007). A technique for representing multiple output binary functions with applications to verification and simulation. *IEEE Transactions on Computers*, **56**(8):1133–1145.

[5] G. DeMicheli (1994). *Synthesis and Optimization of Digital Circuits*. McGraw-Hill.

[6] R. Anderson (1997). An Introduction to Binary Decision Diagrams, www.itu.dk/people/hra/notes-index.html.

[7] G. Hachtel and F. Somenzi (1998). *Logic Synthesis and Verification Algorithms*. Kluwer Academic.

[8] R. Rudell (1993). Dynamic variable ordering for ordered binary decision diagrams. In *IEEE International Conference on Computer-Aided Design*, pp. 47–42.

[9] O. Coudert and J. C. Madre (1990). A unified framework for the formal verification of sequential circuits. In *Proceedings of ICCAD*, pp. 126–129.

[10] H. J. Touati, H. Savoj, B. Lin, R. K. Brayton, and A. Sangiovanni-Vincentelli (1990). Implicit state enumeration of finite state machines using BDDs. In *Proceedings of ICCAD*, pp. 130–133.

[11] E. A. Emerson (1990). Temporal and modal logic. In J. van Leeuwen, ed., *Formal Models and Semantics*, Handbook of Theoretical Computer Science, vol. B, pp. 996–1072. Elsevier Science.

[12] K. L. McMillan (1993). *Symbolic Model Checking*. Kluwer Academic.

[13] R. E. Bryant and Y.-A. Chen (1995). Verification of arithmetic functions with binary moment diagrams. In *Proceedings of the Design Automation Conference*, pp. 535–541.

[14] K. S. Brace, R. Rudell, and R. E. Bryant (1990). Efficient implementation of the BDD package. In *Proceedings of the Design Automation Conference*, pp. 40–45.

[15] F. Somenzi (1998). *CUDD: CU Decision Diagram Package*. Release 2.3.0, University of Colorado at Boulder, USA. http://vlsi.colorado.edu/~fabio/CUDD/.

[16] R. K. Brayton, G. D. Hachtel, A. Sangiovanni-Vencentelli, *et al.* (1996). Vis: a system for verification and synthesis. In *Proceedings of the Computer Aided Verification Conference*, pp. 428–432.

[17] U. Kebschull, E. Schubert, and W. Rosentiel (1992). Multilevel logic synthesis based on functional decision diagrams. In *EDAC*, pp. 43–47.

[18] R. Drechsler, A. Sarabi, M. Theobald, B. Becker, and M. A. Perkowski (1994). Efficient representation and manipulation of switching functions based on order kronecker function decision diagrams. In *Proceedings of the Design Automation Conference*, pp. 415–419.

[19] S. Minato (1993). Zero-suppressed BDDS for set manipulation in combinatorial problems. In *Proceedings of the Design Automation Conference*, pp. 272–277.

[20] R. E. Bryant (1995). Binary decision diagrams and beyond: enabling technologies for formal verification. In *International Conference on Computer-Aided Design*.

[21] S. Höreth and R. Drechsler (1999). Formal verification of word-level specifications. In *Design, Automation and Test in Europe*, pp. 52–58.

[22] E. M. Clarke, K. L. McMillan, X. Zhao, M. Fujita, and J. Yang (1993). Spectral transforms for large boolean functions with applications to technology mapping. In *Proceedings of the Design Automation Conference*, pp. 54–60.

[23] I. Bahar, E. A. Frohm, C. M. Gaona, *et al.* (1993). Algebraic decision diagrams and their applications. In *International Conference on Computer Aided Design*, pp. 188–191.

[24] Y-T. Lai and S. Sastry (1992). Edge-valued binary decision diagrams for multi-level hierarchical verification. In *Proceedings of the Design Automation Conference*, pp. 608–613.

[25] Y-T. Lai, M. Pedram, and S. B. Vrudhula (1993). FGILP: an ILP solver based on function graphs. In *International Conference on Computer-Aided Design*, pp. 689–685.

[26] Y-A. Chen and R. Bryant (1997). PHDD: an efficient graph representation for floating point circuit verification. In *IEEE International Conference on Computer-Aided Design*, pp. 2–7.

[27] R. Drechsler, B. Becker, and S. Ruppertz (1997). The K*BMD: a verification data structure. *IEEE Design & Test of Computers*, **14**(2):51–59.

[28] H. Enderton (1972). *A Mathematical Introduction to Logic*. Academic Press.

[29] T. Bultan, R. Gerber, and C. League (1998). Verifying systems with integer constraints and Boolean predicates: a composite approach. In *Proceedings of the International Symposium on Software Testing and Analysis*, pp. 113–123.

[30] S. Devadas, K. Keutzer, and A. Krishnakumar (1991). Design verification and reachability analysis using algebraic manipulation. In *Proceedings of the International Conference on Computer Design*.

[31] G. Ritter (1999). Formal verification of designs with complex control by symbolic simulation. In *Advanced Research Working Conference on Correct Hardware Design and Verification Methods (CHARME)*. Springer-Verlag LCNS.

[32] R. E. Shostak (1984). Deciding combinations of theories. *Journal of ACM*, **31**(1):1–12.

[33] A. Stump, C. W. Barrett, and D. L. Dill (2002). CVC: a cooperating validity checker. In E. Brinksma and K. Guldstrand Larsen, eds., *14th International Conference on Computer Aided Verification (CAV)*, Lecture Notes in Computer Science, vol. 2404, pp. 500–504. Springer-Verlag.

[34] M. Chandrashekhar, J. P. Privitera, and J. W. Condradt (1987). Application of term rewriting techniques to hardware design verification. In *Proceedings of the Design Automation Conference*, pp. 277–282.

[35] Z. Zhou and W. Burleson (1995). Equivalence checking of datapaths based on canonical arithmetic expressions. In *Proceedings of the Design Automation Conference*.

[36] S. Vasudevan (2003). *Automatic Verification of Arithmetic Circuits in RTL using term rewriting systems*. M.S. thesis, University of Texas, Austin.

[37] J. Burch and D. Dill (1994). Automatic verification of pipelined microprocessor control. In *Computer Aided Verification*. LCNS, Springer-Verlag.

[38] R. Bryant, S. German, and M. Velev (2001). Processor verification using efficient reductions of the logic of uninterpreted functions to propositional logic. *ACM Transactions in Computational Logic*, **2**(1):1–41.

[39] M. Velev and R. Bryant (2003). Effective use of Boolean satisfiability procedures in the formal verification of superscalar and VLIW Microprocessors. *Journal of Symbolic Computation*, **35**(2):73–106.

[40] R. Bryant, S. Lahiri, and S. Seshia (2002). Modeling and verifying systems using a logic of counter arithmetic with lambda expressions and uninterpreted functions. In D. Brinksma and K. G. Larsen, eds., *Computer Aided verification*, Lecture Notes in Computer Science, vol. 2404, pp. 106–122. Springer.

[41] A. Goel, K. Sajid, H. Zhou, A. Aziz, and V. Singhal (2003). BDD based procedures for a theory of equality with uninterpreted functions. *Formal Methods in System Design*, **22**(3):205–224.

[42] L. Arditi (1996). *BMDs can delay the use of theorem proving for verifying arithmetic assembly instructions. In *Proceedings of Formal Methods in CAD (FMCAD)*. Springer-Verlag.

[43] M. Moskewicz, C. Madigan, L. Zhang, Y. Zhao, and S. Malik (2001). Chaff: engineering an efficient SAT solver. In *Proceedings of the 38th Design Automation Conference*, pp. 530–535.

[44] E. Goldberg and Y. Novikov (2002). BerkMin: a fast and robust SAT-solver. In *Proceedings of Design Automation and Test in Europe, DATE-02*, pp. 142–149.

[45] C.-Y. Huang and K.-T. Cheng (2001). Using word-level ATPG and modular arithmetic constraint solving techniques for assertion property checking. *IEEE Transactions on Computer-Aided Design of Integrated Circuits and Systems*, **20**:381–391.

[46] M. Iyer (2003). RACE: a word-level ATPG-based constraints solver system for smart random simulation. In *International Test Conference, ITC-03*, pp. 299–308.

[47] R. Brinkmann and R. Drechsler (2002). RTL-datapath verification using integer linear programming. In *Proceedings of ASP-DAC*.

[48] Z. Zeng, P. Kalla, and M. Ciesielski (2001). LPSAT: a unified approach to rtl satisfiability. In *Proceedings DATE*, pp. 398–402.

[49] F. Fallah, S. Devadas, and K. Keutzer (1998). Functional vector generation for HDL models using linear programming and 3-satisfiability. In *Proceedings of the Design Automation Conference*, pp. 528–533.

[50] G. Bioul and M. Davio (1972). Taylor expansion of Boolean functions and of their derivatives. *Philips Research Reports*, **27**(1):1–6.

[51] A. Thayse and M. Davio (1973). Boolean differential calculus and its application to switching theory. *IEEE Transactions on Computers*, **C-22**(4):409–420.

[52] Maple. www.maplesoft.com.

[53] Mathematica. www.wolfram.com.

[54] The MathWorks. *Matlab*. www.mathworks.com.

[55] M. Ganai, L. Zhang, P. Ashar, A. Gupta, and S. Malik (2002). Combining strengths of circuit-based and CNF-based algorithms for a high-performance SAT solver. In *Design Automation Conference (DAC-2002)*, pp. 747–750.

[56] Z. Zeng, K. Talupuru, and M. Ciesielski (2005). Functional test generation based on word-level SAT. In *Journal of Systems Architecture*, **5**:488–511.

[57] R. E. Bryant and Y-A. Chen (1995). Verification of arithmetic functions with binary moment diagrams. In *Design Automation Conference*, pp. 535–541.

[58] Y. A. Chen and R. E. Bryant (1997). *PHDD: an efficient graph representation for floating point verification. In *Proceedings of the International Conference on Computer Aided Design*.

[59] D. Stoffel and W. Kunz (2004). Equivalence checking of arithmetic circuits on the arithmetic bit level. *IEEE Transactions on CAD*, **23**(5):586–597.

[60] N. Shekhar, P. Kalla, F. Enescu, and S. Gopalakrishnan (2005). Equivalence verification of polynomial datapaths with fixed-size bit-vectors using finite ring algebra. In *International Conference on Computer-Aided Design*.

[61] P. Sanchez and S. Dey (1999). Simulation-based system-level verification using polynomials. In *High-Level Design Validation and Test Workshop, HLDVT*.

[62] R. Drechsler (2000). *Formal Verification of Circuits*. Kluwer Academic.

[63] Y. Lu, A. Koelbl, and A. Mathur (2005). Formal equivalence checking between system-level models and RTL, embedded tutorial. In *International Conference on Computer Aided Design (ICCAD'05)*.

[64] P. Georgelin and V. Krishnaswamy (2006). Towards a C++-based design methodology facilitating sequential equivalence checking. In *Design Automation Conference (DAC'06)*, pp. 93–96.

[65] D. Brier and R. S. Mitra (2006). Use of C/C++ models for architecture exploration and verification of DSPs. In *Design Automation Conference (DAC'06)*, pp. 79–84.

[66] Calypto Design Systems. www.calypto.com.

[67] A. Vellelunga and D. Giramma (2004). The formality equivalence checker provides industry's best arithmetic verification coverage. *Verification Avenue, Synopsys Technical Bulletin*, **5**(2):5–9.

[68] E. Kryrszig (1999). *Advanced Engineering Mathematics*. John Wiley and Sons, Inc.

[69] F. Winkler (1996). *Polynomial Algorithms in Computer Algebra*. Springer.

[70] M. Ciesielski, P. Kalla, Z. Zeng, and B. Rouzeyre (2002). Taylor expansion diagrams: a compact canonical representation with applications to symbolic verification. In *Design Automation and Test in Europe*, pp. 285–289.

[71] P. Coussy and D. Heller (2006). *GAUTY – High-Level Synthesis Tool From C to RTL*. Université de Bretagne-Sud. www-labsticc.univ-ubs.fr/www-gaut/.

[72] R. Rudell (1993). Dynamic variable ordering for binary decision diagrams. In *Proceedings of the International Conference on Computer-Aided Design*, pp. 42–47.

[73] D. Gomez-Prado, Q. Ren, S. Askar, M. Ciesielski, and E. Boutillon (2004). Variable ordering for Taylor expansion diagrams. In *IEEE International High Level Design Validation and Test Workshop, HLDVT-04*, pp. 55–59.

[74] M. Ciesielski, S. Askar, D. Gomez-Prado, J. Guillot, and E. Boutillon (2007). Data-flow transformations using Taylor expansion diagrams. In *Design Automation and Test in Europe*, pp. 455–460.

[75] P. Jain (2002). *Parameterized Motion Estimation Architecture for Dynamically Varying Power and Compression Requirements*. M.S. thesis, Dept. of Electrical and Computer Engineering, University of Massachusetts.

[76] D. Pradhan, S. Askar, and M. Ciesielski (2003). Mathematical framework for representing discrete functions as word-level polynomials. In *IEEE International High Level Design Validation and Test Workshop, HLDVT-03*, pp. 135–139.

[77] W. Stallings (1999). *Cryptography and Network Security*. Prentice Hall.

[78] S. B. Wicker (1995). *Error Control Systems for Digital Communication and Storage*. Prentice Hall.

[79] R. E. Blahut (1984). *Fast Algorithms for Digital Signal Processing*. Addison-Wesley.

[80] T. Kam, T. Villa, R. K. Brayton, and A. L. Sangiovanni-Vincentelli (1998). Multi-valued decision diagrams: theory and applications. *Multiple Valued Logic*, **4**(1–2):9–62.

[81] D. M. Miller and R. Drechsler (2002). On the construction of multiple-valued decision diagrams. In *Proceedings of the 32nd ISMVL*, pp. 245–253.

[82] C. Scholl, R. Drechsler, and B. Becker (1997). Functional simulation using binary decision diagrams. In *International Conference Computer-Aided Design (ICCAD'97)*, pp. 8–12.

[83] Y. Jiang and R. K. Brayton (2002). Software synthesis from synchronous specification using logic simulation techniques. In *Design Automation Conference (DAC'02)*, pp. 319–324.

[84] D. K. Pradhan (1978). A theory of Galois switching functions. *IEEE Transactions on Computers*, **C–27**(3):239–249.

[85] K. M. Dill, K. Ganguly, R. J. Safranek, and M. A. Perkowski (1997). A new Zhegalkin Galois logic. In *Proceedings of the Read-Müller–97 Conference*, pp. 247–257.

[86] C. H. Wu, C. M. Wu, M. D. Sheih, and Y. T. Hwang (2004). High-speed, low-complexity systolic design of novel iterative division algorithm in GF(2^m). *IEEE Transactions on Computers*, **53**:375–380.

[87] P. C. McGeer, K. L. McMillan, and A. L. Sangiovanni-Vincentell (1995). Fast discrete function evaluation using decision diagram. In *International Conference Computer-Aided Design (ICCAD'95)*, pp. 402–407.

[88] M. J. Ciesielski, P. Kalla, Z. Zeng, and B. Rouzeyere (2002). Taylor expansion diagrams: a compact, canonical representation with applications to symbolic verification. In *Design Automation and Test in Europe*.

[89] A. Jabir and D. Pradhan (2004). MODD: a new decision diagram and representation for multiple output binary functions. In *Design Automation and Test in Europe (DATE'04)*, pp. 1388–1389.

[90] R. S. Stankovi and R. Drechsler (1997). Circuit design from Kronecker Galois field decision diagrams for multiple-valued functions. In *ISMVL-27*, pp. 275–280.

[91] D. K. Pradhan, M. Ciesielski, and S. Askar (2003). Mathematical framework for representing discrete functions as word-level polynomials. In *Proceedings of the HLDVT'03*, pp. 135–142.

[92] P. Asher and S. Malik (1995). Fast functional simulation using branching programmes. In *ICCAD'95*, pp. 408–412.

[93] J. T. Butler, T. Sasao, and M. Matsuura (2005). Average path length of binary decision diagrams. *IEEE Transaction Computers*, **54**(9):1041–1053.

[94] T. Sasao, Y. Iguchi, and M. Matsuura (2002). Comparison of decision diagrams for multiple-output logic functions. In *Proceedings of the IWLS*.

[95] A. Reyhani-Masoleh and M. A. Hasan (2004). Low complexity bit parallel architectures for polynomial basis multiplication over GF(2^m). *IEEE Transactions on Computers*, **53**(8):945–959.

[96] S. Minato (1993). Zero-suppressed BDDs for set manipulation in combinatorial problems. In *Proceedings of the 30th IEEE/ACM Design Automation Conference (DAC'93)*, pp. 272–277.

[97] A. Jabir, T. Rajaprabhu, D. Pradhan, and A. Singh (2004). MODD for CF: a compact representation for multiple-output functions. In *Proceedings of the International Conference High Level Design and the Value Testing (HLDVT'04)*.

[98] F. Schmiedle, W. Gunther, and R. Drechsler (2001). Selection of efficient re-ordering heuristics for MDD construction. In *Proceedings of the 31st IEEE International Symposium on Multi-Valued Logic (ISMVL'01)*, pp. 299–304.

8 Boolean satisfiability and EDA applications

Joao Marques-Silva

8.1 Introduction

Boolean satisfiability (SAT) is a widely used modeling framework for solving combinatorial problems. It is also a well-known decision problem in theoretical computer science, being the first problem to be shown to be NP-complete. [11] Since SAT is NP-complete, and unless $P = NP$, all SAT algorithms require worst-case exponential time. However, modern SAT algorithms are extremely effective at coping with large search spaces, by exploiting the problem's structure when it exists. [2–4] The performance improvements made to SAT solvers since the mid 1990s motivated their application to a wide range of practical applications, from cross-talk noise prediction in integrated circuits [5] to termination analysis in term-rewrite systems. [6] In some applications, the use of SAT provides remarkable performance improvements. Examples include model-checking of finite-state systems, [7–9] design debugging, [10] AI planning, [11,12] and haplotype inference in bioinformatics. [13] Additional successful examples of practical applications of SAT include termination analysis in term-rewrite systems, [6] knowledge-compilation, [4] software-model checking, [15,16] software testing, [17] package management in software distributions, [18] checking of pedigree consistency, [19] verification of pipelined processors, [20-21] symbolic-trajectory evaluation, [22] test-pattern generation in digital systems, [23] design debugging and diagnosis, [10] identification of functional dependencies in Boolean functions, [24] technology-mapping in logic synthesis, [25] circuit-delay computation, [26] and cross-talk-noise prediction. [5] However, this list is incomplete, as the number of applications of SAT has been on the rise in recent years. [18,19,24]

Besides practical applications, SAT has also influenced a number of related decision and optimization problems, which will be referred to as extensions of SAT. Most extensions of SAT either use the same algorithmic techniques as used in SAT, or use SAT as a core engine. One of the most promising extensions of SAT is satisfiability modulo theories (SMT). [27,28] Other applications of SAT include pseudo-Boolean (PB) constraints, [29,30] maximum satisfiability (MaxSAT), [31,32] model counting (#SAT), [33,34] and quantified-Boolean formulas (QBF). [35]

Practical Design Verification, eds. Dhiraj K. Pradhan and Ian G. Harris. Published by Cambridge University Press. © Cambridge University Press 2009.

As illustrated, many practical applications of SAT are in electronic design automation (EDA) and in the areas of verification, [7–9,20,22,26] testing, [23,36,37] and also synthesis. [24,25] Accordingly, this chapter overviews both SAT algorithms and some of the most successful practical applications of SAT in EDA. Moreover, the chapter summarizes some other well-known applications, and overviews the use of SAT in some of its best-known extensions. The chapter is organized as follows. Section 8.2 introduces the notation used in the remainder of the chapter. Section 8.3 outlines research work in representative extensions of SAT. Afterwards, Section 8.4 illustrates practical applications of SAT, by focusing on a number of concrete case studies. Finally, the chapter concludes in Section 8.5.

8.2 Definitions

8.2.1 Propositional formulas and satisfiability

Propositional formulas are defined over a finite set of Boolean variables X. Individual variables can be represented by letters x, y, z, w, and o, and subscripts may be used (e.g., x_1). The propositional connectives considered will be $\neg$, $\vee$, $\wedge$, $\rightarrow$, and $\leftrightarrow$. Parentheses will be used to enforce precedence. Most SAT algorithms require propositional formulas to be represented in conjunctive normal form (CNF). A CNF formula φ consists of a conjunction of clauses ω, each of which consists of a disjunction of literals. A literal is either a variable x_i or its complement $\neg x_i$. A CNF formula can also be viewed as a set of clauses, and each clause can be viewed as a set of literals. Throughout this chapter, the representation used will be clear from the context. Consider a CNF formula φ with two clauses ω_j and ω_k, such that $\omega_j \subseteq \omega_k$. Then ω_j is said to *subsume* ω_k. All variable assignments that satisfy ω_j also satisfy ω_k. In this situation, ω_k can be removed from φ. Arbitrary propositional formulas can be converted to CNF in linear time and space by adding additional variables. This conversion is addressed in the next section.

In the context of search algorithms for SAT, variables can be *assigned* a logic value, either 0 or 1. Alternatively, variables may also be *unassigned*. Assignments to the problem variables can be defined as a function $v : X \rightarrow \{0, u, 1\}$, where u denotes an *undefined* value used when a variable has not been assigned a value in $\{0,1\}$. Given an assignment v, if all variables are assigned a value in $\{0,1\}$, then v is referred to as a *complete assignment*. Otherwise it is a *partial assignment*.

Assignments serve for computing the values of literals, clauses, and the complete CNF formula, respectively, l^v, ω^v, and φ^v. A total order is defined on the possible assignments, $0 < u < 1$. Moreover, $1 - u = u$. As a result, the following definitions apply:

$$l^v = \begin{cases} v(x_i) & \text{if } l = x_i \\ 1 - v(x_i) & \text{if } l = \neg x_i, \end{cases} \tag{8.1}$$

$$\omega^v = \max\{l^v | l \in \omega\}, \tag{8.2}$$

$$\varphi^v = \min\{\omega^v | \omega \in \varphi\}. \tag{8.3}$$

The assignment function v will also be viewed as a set of tuples (x_i, v_i), with $v_i \in \{0, 1\}$. Adding a tuple (x_i, v_i) to v corresponds to assigning v_i to x_i, such that $v(x_i) = v_i$. Removing a tuple (x_i, v_i) from v, with $v(x_i) \neq u$, corresponds to assigning u to x_i.

Given an assignment, clauses and CNF formulas can be characterized as *unsatisfied*, *satisfied*, or *unresolved*. A clause is unsatisfied if all its literals are assigned value 0. A clause is satisfied if at least one of its literals is assigned value 1. A clause is unresolved if it is neither unsatisfied nor satisfied. A CNF formula φ is satisfied if *all* clauses are satisfied, and is unsatisfied if at least one clause is unsatisfied. Otherwise it is unresolved. The SAT problem for a CNF formula φ consists of deciding whether there exists an assignment to the problem variables, such that a given CNF formula φ is satisfied, or proving that no such assignment exists. As mentioned earlier, the satisfiability problem for general propositional formulas is NP-complete [1] and so is the satisfiability problem for CNF formulas.

Given a partial assignment, an unresolved clause such that all but one literal are assigned value 0, and the remaining literal is assigned, is said to be *unit*. [38] A key procedure in SAT algorithms is the *unitary clause rule*: [38] if a clause is unitary, then its sole unassigned literal must be assigned value 1 for the clause to be satisfied. The iterated application of the unit clause rule is referred to as *unit propagation* or *Boolean constraint propagation* (BCP). [39]

8.2.1.1   Resolution

Resolution [38] represents a fundamental operation in Boolean satisfiability. Let $\odot$ represent the resolution operator. For two clauses ω_j and ω_k, for which there is a unique variable x such that one clause has a literal x and the other has literal $\neg x$, $\omega_j \odot \omega_k$ contains all the literals of ω_j and ω_k with the exception of x and $\neg x$.

Resolution forms the basis of one of the first algorithms for SAT [38] and is an often-used technique for preprocessing CNF formulas. [40,41] Even though resolution-based algorithms are not effective in practice, modern SAT algorithms use the resolution operation in a number of ways, including during preprocessing and for learning new clauses. These uses are described below.

Boolean satisfiability algorithms based on resolution iteratively remove variables by applying the resolution operation between all pairs of clauses containing a literal and its complement. Satisfied clauses are discarded, and the process stops when no more resolution operations can be performed. If the empty clause is derived, the original CNF formula is unsatisfiable.

Example 8.1 Consider the CNF formula:

$$\varphi = (x_1 \vee x_2) \wedge (x_1 \vee \neg x_3) \wedge (\neg x_1 \vee \neg x_2) \wedge (\neg x_1 \vee \neg x_3).$$

By applying the resolution operation for removing x_1, the following CNF formula is obtained:

$$\varphi' = (x_2 \vee \neg x_2) \wedge (\neg x_2 \vee \neg x_3) \wedge (\neg x_2 \vee \neg x_3) \wedge (\neg x_3).$$

Table 8.1 CNF representation of simple gates

Gate	CNF representation
$y = \text{NOT}(x1)$	$(\neg y \vee \neg x_1)(y \vee x_1)$
$y = \text{AND}(x_1, \ldots, x_k)$	$(y \vee \neg x_1 \neg \ldots \vee \neg x_k) \wedge \Lambda_{i=1}^{k}(x_i \vee \neg y)$
$y = \text{OR}(x_1, \ldots, x_k)$	$(y \vee \neg x_1 \neg \ldots \vee \neg x_k) \wedge \Lambda_{i=1}^{k}(x_i \vee \neg y)$

The first clause is trivially satisfied and can be removed. The second and third clauses are subsumed by the fourth clause, and so can also be removed. Hence, the resulting CNF formula becomes:

$$\varphi'' = (\neg x_3).$$

No more resolution operations can be applied. The empty clause was not derived. Hence the formula is satisfiable, $x_3 = 0$ is a necessary assignment. The remaining assignments can be identified by branching and propagation of necessary assignments.

8.2.2 Boolean circuits

Many practical applications are often represented in some intermediate representation, from which a CNF formula is then generated. Combinational circuits are one of the most often used intermediate representations. [7,11,15,18] Combinational Boolean circuits are composed of gates and connections between gates. In this chapter, only simple gates are considered and restricted to basic operations: NOT, AND, OR, XOR, or alternatively $^{-}$, ., +, $\oplus$. Observe that $\text{XOR}(x,y) = \text{OR}(\text{AND}(x,\text{NOT}(y)), \text{AND}(\text{NOT}(x), y))$, or, alternatively, $x \oplus y = x \cdot \bar{y} + \bar{x} \cdot y$. Moreover, for simplicity, two-input single-output gates are assumed. The notation $y = \text{OP}(x_1, x_2)$ denotes a gate with output y, and inputs x_1 and x_2, and OP is one of the basic operations.

Converting Boolean circuits to CNF is straightforward, and follows the procedure outlined by G. Tseitin. [42] Consider a gate $y = \text{OP}(x_1, x_2)$. The CNF representation captures the valid assignments between the gate inputs and outputs. Hence, $\varphi(y, x_1, x_2) = 1$ if the predicate $y = \text{OP}(x_1, x_2)$ holds true. The CNF representations for simple gates are shown in Table 8.1 (observe that XOR gates can be replaced by NOT, AND, and OR as described above). For generality, the number of inputs considered for AND and OR gates is unrestricted. Even though Tseitin's transformation is arguably the most often used, there are a number of effective alternatives including Plaisted and Greenbaum's. [43]

Example 8.2 For the example circuit of Fig. 8.1, the resulting CNF formula using Tseitin's transformation is:

$$\begin{aligned}
\varphi = &(a \vee x) \wedge (b \vee x) \wedge (\neg a \vee \neg b \vee \neg x) \wedge \\
&(x \vee \neg y) \wedge (c \vee \neg y) \wedge (\neg x \vee \neg c \vee y) \wedge \\
&(\neg y \vee z) \wedge (\neg d \vee z) \wedge (y \vee d \vee \neg z) \wedge (z).
\end{aligned}$$

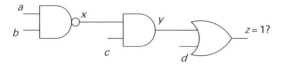

Figure 8.1   Example circuit

Another often-used technique consists of exploiting the sharing of common structure in Boolean circuits. Examples of representations that exploit structural sharing are *reduced Boolean circuits* (RBC), [44] *Boolean expression diagrams* (BED), [45] and *and-inverter graphs* (AIG). [46]

Observe that it is straightforward to represent arbitrary propositional formulas as Boolean circuits. First, note that $\neg$, $\wedge$, and $\vee$ represent a sufficient set of connectives. Second, associate a new Boolean variable with each level of parenthesis in the propositional formula. As a result, it is straightforward to represent arbitrary propositional formulas in CNF.

8.2.3   Linear inequalities over Boolean variables

Linear inequalities over Boolean variables are a widely used modeling technique. For example, with the objective of modeling an integer variable r that can take one out of k values, i.e., $1 \leq r \leq k$, one often-used approach is to create k Boolean variables $x_1, \ldots, x_k$, such that $x_i = 1$, $1 \leq i \leq k$, if $r = i$. In addition, since r must take one of its possible values, then one of the x_i variables must be assigned value 1. Hence,

$$\sum_{i=1}^{k} x_{ij} = 1, \tag{8.4}$$

which can be represented as:

$$\left(\sum_{i=1}^{k} x_i \leq 1 \right) \wedge \left(\sum_{i=1}^{k} x_i \geq 1 \right). \tag{8.5}$$

The previous example illustrates special cases of linear inequalities, referred to as *cardinality constraints*, the general form being $\Sigma\, x_i \leq k$. More general constraints are often necessary, and so in general it is necessary to develop solutions for encoding linear inequalities of the form:

$$\sum_{i=1}^{k} a_i x_i \leq b. \tag{8.6}$$

The encoding proposed by J. Warners [47] ensures that linear inequalities can be encoded into CNF in linear time and space, and uses adders as the basic operator. Despite being optimal in terms of the space required, Warners's encoding does not guarantee *arc-consistency*, i.e., the ability to imply all necessary assignments given a partial assignment. Other encodings exist [30,48] that can use binary decision diagrams (BDDs) or sorting networks, among other structures. For arbitrary linear inequalities, BDDs guarantee arc-consistency but can require exponential space in the worst case. Sorting networks require polynomial space but do not guarantee arc-consistency.

For cardinality constraints, a number of polynomial encodings ensure arc-consistency, including BDDs, sorting networks, [30] and sequential counters. [49] Given its widespread use, the encoding for $\Sigma\, x_i \leq 1$ using sequential counters is:

$$
\begin{aligned}
&(\neg x_1 \vee s_1) \wedge (\neg x_k \vee \neg s_{k-1}) \wedge \\
&\Lambda_{1 < i < k}((\neg x_i \vee s_i) \wedge (\neg s_{i-1} \vee s_i) \wedge (\neg x_i \vee \neg s_{i-1})),
\end{aligned}
\tag{8.7}
$$

where s_i are additional auxiliary Boolean variables. Inspection of the formula allows the conclusion that, at most, one x_i can be assigned value 1, for which s_{i-1}, with $i > 1$, is assigned value 0 and s_i is assigned value 1. For all x_i, with $i > 1$, for which $s_{i-1} = s_i$, then x_i must be assigned value 0. Moreover, observe that encoding $\Sigma\, x_i \geq 1$ is immediate with a single clause and, given Eq. 8.7, so is the encoding of $\Sigma\, x_i = 1$.

An alternative solution, which reduces the number of auxiliary variables is to use *bitwise encoding*. [50] Consider constraint $\Sigma\, x_i \leq 1$. Create r auxiliary variables, where $r = 1$ if $k = 1$ and $r = \lfloor \log k \rfloor$ if $k > 1$. Let $u_0, \ldots, u_{r-1}$ be the auxiliary variables. Now associate with each x_i the binary representation of $i - 1$. Finally, for each x_i create the clauses: $(\neg x_i \vee p_j), j = 0, \ldots, r - 1$, where $p_j = u_j$ if the binary representation of $i - 1$ has value 1 in position j, and $p_j = -u_j$ otherwise. If a given variable x_i is assigned value 1, then the literals p_j in the binary clauses must be assigned value 1, thus encoding the binary representation of $i - 1$. Since the p_j literals can encode, at most, one binary representation, all other x_i variables must be assigned value 0. For constraint $\Sigma\, x_i \leq 1$ with k variables, the bitwise encoding requires $O(\log k)$ variables and $O(k \log k)$ clauses, i.e., $O(\log k)$ for each variable in the constraint.

Finally, more general constraints can be encoded into CNF (e.g., [51]), albeit this is seldom used in practical settings.

8.2.4 SAT algorithms

A vast number of different algorithms have been proposed for the SAT problem over the years. [52] Examples include different proof systems, [38,53] backtrack searching, [54] and local searching. [55] In addition, dedicated solvers have been developed for non-clausal forms, including, for example, automatic test pattern generation (ATPG) algorithms [56,57] and recursive learning. [58]

Despite the existence of many alternative algorithms for SAT, the most effective for solving satisfiability in EDA problems are based on backtrack searching with clause learning. These algorithms are referred to as *conflict-driven clause learning* (CDCL) SAT solvers, and are overviewed in the next section.

In addition, the following sections summarize recent work on non-clausal SAT solvers, and also techniques for preprocessing instances of SAT, both in clausal and in non-clausal form.

8.2.4.1   CDCL SAT algorithms

The CDCL SAT solvers are derived from the well-known DPLL SAT algorithm, first described in [54], but including techniques first proposed in [38]. Moreover, CDCL SAT solvers implement a number of essential search techniques, including clause learning and non-chronological backtracking. [2]

In modern CDCL solvers, as in most implementations of DPLL, logical consequences are derived with unit propagation. Unit propagation is applied after each branching step (and also during preprocessing), and is used for identifying variables that must be assigned a specific Boolean value. If an unsatisfied clause is identified, a *conflict* condition is declared, and the algorithm backtracks.

In CDCL SAT solvers, each variable x_i is characterized by a number of properties, including the *value*, the *antecedent*, and the *decision level*, denoted respectively by $v(u_i) \in \{0, u, 1\}$, $a(x_i) \in \varphi \cup \{\text{NIL}\}$, and $\delta(x_i) \in \{-1, 0, 1, \ldots, |X|\}$. A variable x_i that is assigned a value as the result of applying the unit clause rule is said to be *implied*. The unit clause ω used for implying variable x_i is said to be the antecedent of x_i, $a(x_i) = \omega$. For variables that are decision variables or are unassigned, the antecedent is NIL. Hence, antecedents are only defined for variables whose value is implied by other assignments. The decision level of a variable x_i denotes the depth of the decision tree at which the variable is assigned a value in $\{0,1\}$. The decision level for an unassigned variable x_i is -1, $\delta(x_i) = -1$. The decision level associated with variables used for branching steps (i.e., *decision assignments*) is specified by the search process, and denotes the current depth of the *decision stack*. Hence, a variable x_i associated with a decision assignment is characterized by having $a(x_i) = \text{NIL}$ and $\delta(x_i) > 0$. Alternatively, the decision level of x_i with antecedent ω is given by:

$$\delta(x_i) = \max(\{0\} \cup \{\delta(x_j)|x_j \in \omega \wedge x_j \neq x_i\}). \tag{8.8}$$

The notation $x_i = v @ d$ is used to denote that $v(x_i) = v$ and $\delta(x_i) = d$. Moreover, the decision level of a literal is defined as the decision level of its variable, $\delta(l) = \delta(x_i)$ if $l = x_i$ or $l = \neg x_i$.

During the execution of a DPLL-style SAT solver, assigned variables, as well as their antecedents, define a directed acyclic graph $I = (V_I, E_I)$, referred to as the *implication graph*. [2]

The vertices in the implication graph are defined by all assigned variables and one special node κ, $V_I \subseteq X \cup \{\kappa\}$. The edges in the implication graph are obtained from the antecedent of each assigned variable: if $\omega = a(x_i)$, then there is a directed edge from each variable in ω, other than x_i, to x_i. If unit propagation yields an unsatisfied clause ω_j, then a special vertex κ is used to represent the unsatisfied clause. In this case, the antecedent of κ is defined by $a(\kappa) = \omega_j$.

Example 8.3 (Implication graph) Consider the CNF formula:

$$\begin{aligned}
\varphi_1 &= \omega_1 \wedge \omega_2 \wedge \omega_3 \wedge \omega_4 \wedge \omega_5 \wedge \omega_6 \\
&= (x_1 \vee x_{31} \vee \neg x_2) \wedge (x_1 \vee \neg x_3) \wedge (x_2 \vee x_3 \vee x_4) \\
&\quad \wedge (\neg x_4 \vee \neg x_5) \wedge (x_{21} \vee \neg x_4 \vee \neg x_6) \wedge (x_5 \vee x_6).
\end{aligned} \tag{8.9}$$

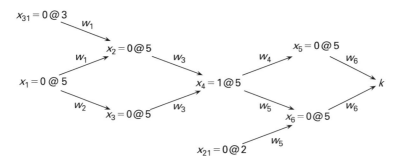

Figure 8.2   Implication graph for Example 8.3

Assume decision assignments $x_{21} = 0@2$ and $x_{31} = 0@3$. Moreover, assume the current decision assignment $x_1 = 0@5$. The resulting implication graph is shown in Fig. 8.2, and yields a conflict because clause $(x_5 \vee x_6)$ becomes unsatisfied.

In the presence of conflicts, modern CDCL SAT solvers learn new clauses. [2] Learnt clauses are then used to implement non-chronological backtracking. [2]

Example 8.4 (Clause learning) For the CNF formula of Example 8.3, a new clause $(x_1 \vee x_{31} \vee x_{21})$ is learnt by analyzing the causes of the conflict. [2] For this example, the conflict takes place at decision level 5. The analysis of the conflict discards all assignments at decision level 5, with the exception of the decision assignment (i.e., $x_1 = 0$). In addition, all assignments at decision levels less than 5, that are used for deriving the conflict, are also kept. This process can be implemented by traversing the implication graph, with the restriction that only vertices assigned at the current decision level are visited.

Moreover, the structure of the conflicts can be exploited by identifying unique implication points (UIPs). [2] For this example, $x_4 = 1@5$ is a UIP, and so the learnt clause would be $\vee(\neg x_4 \vee x_{21})$. This would be the clause learnt by recent CDCL SAT solvers. [3,4]

Algorithm 8.1 shows the standard organization of a CDCL SAT solver, which essentially follows the organization of DPLL. With respect to DPLL, the main differences are the call to function *Conflict Analysis* each time a conflict is identified, and the call to *Backtrack* when backtracking takes place. Moreover, the Backtrack procedure allows for non-chronological backtracking.

In addition to the main CDCL function, the following auxiliary functions are used:

- *UnitPropagation* consists of the iterated application of the unit clause rule. If an unsatisfied clause is identified, then a conflict indication is returned.
- *PickBranchingVariable* consists of selecting a variable to assign and the respective value.

Algorithm 8.1 Typical CDCL algorithm

CDCL(φ, v)

1 **if** (UNITPROPAGATION(φ, v) == **CONFLICT**)
2   **then return UNSAT**
3 $dl \leftarrow 0$ $\triangleright$ Decision level
4 **while** (**not** ALLVARIABLESASSIGNED(φ, v))
5   **do** $(x, v) =$ PICKBRANCHINGVARIABLE(φ, v) $\triangleright$ DECIDE STAGE
6          $dl \leftarrow dl + 1$ $\triangleright$ New decision: update decision level
7          $v \leftarrow v \cup \{(x, v)\}$
8          $\triangleright$ DEDUCE stage
9   **if** (UNITPROPAGATION(φ, v) == **CONFLICT**)
10   **then** $\beta = $ CONFLICTANALYSIS(φ, v) $\triangleright$ DIAGNOSE stage
11   **if** ($\beta < 0$)
12   **then return UNSAT**
13   **else** Backtrack(φ, v, β)
14                         $dl \leftarrow \beta \triangleright$ Backtracking: update decision level
15 **return SAT**

- *ConflictAnalysis* consists of analyzing the most recent conflict and learning a new clause from the conflict. The organization of this procedure is described elsewhere. [2]
- *Backtrack* backtracks to the decision level computed by *ConflictAnalysis*.
- *AllVariablesAssigned* tests whether all variables have been assigned, in which case the algorithm terminates, indicating that the CNF formula is satisfiable. An alternative criterion to stop execution of the algorithm is to check whether all clauses are satisfied. However, in modern SAT solvers that use lazy data structures, clause state cannot be maintained accurately, and so the termination criterion must be whether all variables are assigned.

Arguments to the auxiliary functions are assumed to be passed by reference. Hence, φ and v are supposed to be modified during execution of the auxiliary functions.

The typical CDCL algorithm shown does not account for a few often-used techniques as well as key implementation details. A state-of-the-art SAT solver implements the typical CDCL algorithm shown above, and also uses the following techniques:

- Identification of unique implication points (UIPs) [2] (see Example 8.4). Unique implication points represent dominators [59] in the implication graph. Given the special structure of implication graphs, UIPs are identified in linear time.
- Memory efficient lazy data structures. [3] Lazy data structures require essentially no effort during backtracking. Moreover, during propagation, only a fraction of a variable's clauses are updated.

- Adaptive branching heuristics, usually derived from the variable state independent decaying sum (VSIDS) heuristic. [3] The VSIDS heuristic associates a weight with each variable. The weights are regularly divided by a constant, and each is incremented when the variable participates in a conflict.
- Integration of search restarts, by using some completeness criterion. [60,61] An often-used completeness criterion is to increase the number of conflicts in between restarts.
- Implementation of clause-deletion policies. [62] Existing clause-deletion policies evaluate how often learnt clauses are used for identifying conflicts. Clauses that are used less often can be deleted.

Because modern backtrack-search SAT solvers learn clauses, it is straightforward to track all the learnt clauses, and use these clauses to construct a resolution refutation (or unsatisfiability proof) of the original formula. [63]

8.2.4.2   Non-clausal SAT algorithms

A number of alternatives to clausal CDCL SAT solvers have been proposed in recent years. [46,57,64–66] Modern non-clausal SAT solvers implement the most effective techniques used in clausal SAT solvers, including clause learning and non-chronological backtracking. In addition, non-clausal SAT solvers use dedicated representations of Boolean networks, e.g., AIG. [46]

One key technique of non-clausal SAT solvers is the identification of shared sub-structures. The existence of shared sub-networks in Boolean circuits allows the reduction of both the number of Boolean variables and the number of clauses used. Another often-used technique is to use structural information to simplify the SAT problem being solved. Examples include maintaining a justification frontier [64] and identifying observability "don't cares." [67] In addition, specialized forms of learning have also been proposed. [46]

8.2.4.3   Preprocessing

Preprocessing of CNF formulas aims at modifying CNF formulas, such that these formulas become simpler to solve by SAT solvers. Preprocessing can consist of adding or removing clauses or variables. [40,41,68–70]

The simplest form of resolution is based on probing value assignments to variables. [70] A number of techniques based on variants of resolution have also been proposed. [40,41,68,69] For example, NiVER [40] applies resolution operations while the number of the literals in the CNF formula can be reduced.

Example 8.5 Consider the CNF formula:

$$(x_1 \vee x_2) \wedge (\neg x_2 \vee x_3) \wedge (\neg x_2 \vee x_4).$$

By applying the resolution operation with respect to x_2, the resulting equivalent CNF formula (after simplification) becomes:

$$(x_1 \vee x_3) \wedge (x_1 \vee x_4).$$

The first formula contains six literals, whereas the second CNF formula contains four literals. Hence, NiVER replaces the original formula by the second one.

8.3   Extensions of SAT

A number of extensions of SAT allow greater modeling flexibility than plain SAT. Purely Boolean examples include quantified Boolean formulas (QBF), pseudo-Boolean (PB) solving and optimization, and maximum satisfiability (MaxSAT) and variants. The most effective algorithmic techniques used in SAT have also been applied in most extensions of SAT, thus enabling significant practical applications. This section briefly surveys these extensions of SAT.

Pseudo-Boolean (PB) constraints generalize SAT by considering linear inequalities over Boolean variables instead of clauses. Moreover, a linear cost function can be considered. The pseudo-Boolean optimization problem can be defined as follows: [71]

$$
\begin{aligned}
&\text{Minimize} \sum_{j \in N} c_j.x_j \\
&\text{subject to} \sum_{j \in N} a_{ij} l_j \geq b_i, \\
&x_j \in \{0,1\}, a_{ij}, b_i \in N_0^+, j \in N, i \in M \\
&N = \{1,...,n\}, M = \{1,...,m\}.
\end{aligned}
\tag{8.10}
$$

The problem of optimizing PB-constraints is NP-hard. Moreover, as in the case of SAT, a number of effective algorithms have been proposed [29,30] that integrate and extend the most effective SAT techniques.

The maximum satisfiability (MaxSAT) problem can be stated as follows. Given an instance of SAT represented in conjunctive normal form (CNF), compute an assignment to the variables that maximizes the number of satisfied clauses. Variations of the MaxSAT problem include the partial MaxSAT problem, the weighted MaxSAT problem, and the weighted partial MaxSAT problem. In the partial MaxSAT problem some clauses (i.e., the *hard* clauses) must be satisfied, whereas others (i.e., the *soft* clauses) may not be satisfied. In the weighted MaxSAT problem, each clause has a given weight, and the objective is to maximize the sum of the weights of satisfied clauses. Finally, in the weighted partial MaxSAT, the hard clauses must be satisfied, a weight is associated with each soft clause, and the objective is to maximize the sum of the weights of satisfied clauses. The MaxSAT problem and its variants provide a versatile modeling solution and a growing number of practical applications, [31,32] including the ability to solve PB optimization problems. Despite the potential applications, the most effective SAT techniques cannot be applied directly in algorithms for MaxSAT. As a result, the best performing algorithms use branch and bound searching with sophisticated bounding. [31,34] Recent work has shown how to use SAT iteratively for solving MaxSAT. [72,73]

One SAT-related decision problem is quantified Boolean formula (QBF), a well-known example of PSPACE-complete decision problems. The QBF problem finds a large number of potential practical applications, including model checking. [74] A QBF formula is a CNF formula where the Boolean variables are quantified, and is of the form:

$$Q_1 x_1 \ Q_2 x_2 \ ... \ Q_n x_n \varphi, \tag{8.11}$$

where $Q_i \in \{\exists, \forall\}$ and φ is a CNF formula. Recent algorithms for QBF have integrated and extended the most effective SAT techniques. [35] Nevertheless, the performance improvements in QBF solvers have not been as significant as in SAT solvers.

Besides extensions of SAT based on Boolean domains, a number of extensions exist, including satisfiability modulo theories. [27,28]

8.4   Applications of SAT in EDA

This section overviews the application of SAT and extensions of SAT in a number of areas, namely combinational equivalence checking, [75] automatic test-pattern generation, [23] design debugging, [76] bounded model checking, [7] and unbounded model checking. [8,9] The applications are organized by increasing problem formulation complexity.

8.4.1   Combinational equivalence checking

An essential circuit design task is to check the functional equivalence of two circuits. The simplest form of equivalence checking addresses combinational circuits. Let C_A and C_B denote two combinational circuits, both with inputs $x_1, \ldots, x_n$ and both with m outputs, C_A with outputs $y_1, \ldots, y_m$ and C_B with outputs $w_1, \ldots, w_m$. The function implemented by each of the two circuits is defined as follows: $\mathbf{f}_A : \{0, 1\}^n \rightarrow \{0, 1\}^m$, and $\mathbf{f}_B : \{0, 1\}^n \rightarrow \{0, 1\}^m$. Let $x \in \{0, 1\}^n$ and define $\mathbf{f}_A(x) = (f_{A,1}(x), \ldots, f_{A,m}(x))$ and $\mathbf{f}_B(x) = (f_{B,1}(x), \ldots, f_{B,m}(x))$. The two circuits are *not* equivalent if the following condition holds:

$$\exists_{x \in \{0,1\}^n} \exists_{1 \leq i \leq m} \ f_{A,i}(x) \neq f_{B,i}(x), \tag{8.12}$$

which can be represented as the following satisfiability problem:

$$\overset{n}{\underset{i=1}{V}} \ (f_{A,i}(x) \oplus f_{B,i}(x)) = 1. \tag{8.13}$$

The resulting satisfiability problem is illustrated in Fig. 8.3, and is referred to as a *miter*. [75] From the results of the previous section it is straightforward to encode the combinational equivalence checking problem in CNF. Somewhat surprisingly, combinational equivalence checking can be challenging for SAT solvers. Hence, a number of techniques, including miter preprocessing and solving intermediate equivalence checking problems, are often used. [46,67]

8.4.2   Automatic test-pattern generation

Fabricated integrated circuits may be subject to defects, which may cause circuit failure. The most widely used approach for identifying fabrication defects is

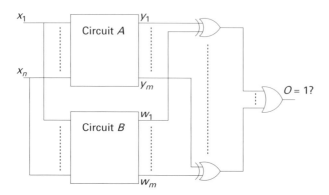

Figure 8.3   Combinational equivalence checking

automatic test-pattern generation (ATPG). [78] Moreover, the most often-used model for representing fabrication defects is the single stuck-at fault model (SSF), [78] where a single connection in the circuit is assumed to be stuck at a given logic value, either 0 or 1, denoted respectively by stuck-at 0 (or sa-0) and stuck-at 1 (or sa-1). Automatic test-pattern generation consists of computing input assignments that allow demonstration of the existence or absence of each target fault, or proof that no assignment exists (hence, it is essentially a modified satisfiability problem). When such an assignment exists, it is said that the target fault has been detected. In what follows, combinational circuits are assumed, but the same ideas can be extended to sequential circuits. [78]

To compute an input assignment to detect a given target fault x sa-v, two copies of the circuit are considered. The first copy represents the circuit without the fault, and is referred to as the *good* circuit. The second copy represents the circuit with the fault, and is referred to as the *faulty* circuit.

Using the notation of the previous section, a Boolean function is associated with each copy of the circuit: the good circuit is described by $\mathbf{f}_G : \{0, 1\}^n \rightarrow \{0, 1\}^m$, and the faulty circuit is described by $\mathbf{f}_F : \{0, 1\}^n \rightarrow \{0, 1\}^m$. As a result, the fault will be detected if for some input assignment, the outputs of the two circuits differ:

$$\exists_{x \in \{0,1\}^n} \exists_{1 \leq i \leq m} f_{G,i}(x) \neq f_{F,i}(x). \tag{8.14}$$

As before, this condition can be represented as the following satisfiability problem:

$$\bigvee_{i=1}^{n} (f_{G,i}(x) \oplus f_{F,i}(x)) = 1. \tag{8.15}$$

Observe that the miter can also be used for representing the problem of ATPG, when A represents the good circuit and B represents the faulty circuit. Even though Eq. 15 can be encoded directly into CNF and solved with a SAT solver, this is, in general, not effective. As a result, the model is modified to provide additional structural infor-mation. [23,36,37] The faulty circuit is only partially represented, involving only the nodes whose value can differ from the good circuit. For each such node x, an

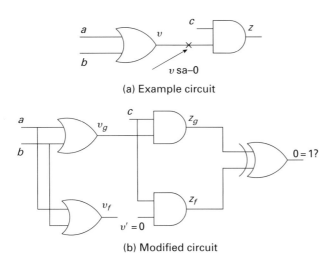

(a) Example circuit

(b) Modified circuit

Figure 8.4   SAT-based ATPG

additional variable x_S is used to denote whether the values in the two circuits differ. The variable x_S is referred to as the sensitization variable of node x and takes value 1 if the values of x in the two circuits differ. If x_G is the value in the good circuit and x_F is the value in the faulty circuit, then x_S is defined as:

$$x_S \leftrightarrow (x_G \oplus x_F). \tag{8.16}$$

Example 8.6 Consider the example circuit in Fig. 8.4(a), with target fault v sa-0. The modified circuit is shown in Fig. 8.4(b). The fault effect is represented by disconnecting u' from u and setting u' to 0. The primary inputs for the two circuits are connected. If there is assignment that satisfies $o = 1$, then the fault is detected. It is straightforward to generate the CNF from the circuit in Fig. 8.4(b), e.g., by using Tseitin's transformation.

The use of SAT in ATPG was first proposed by T. Larrabee. [23] Improvements based on preprocessing were described in [36]. Additional improvements were further proposed in [37], including the reuse of learnt clauses in between target faults and the encoding of conditions for unique sensitization points. [78]

8.4.3   Design debugging

Design debugging is used in the VLSI design cycle to identify design errors at the gate level. Assume a set of input stimuli I and a set of expected output responses O, which can be provided by a simulation tool. The design debugging problem assumes sets I and O and an incorrect circuit C, and seeks to identify gate errors in the design, i.e., gates that implement incorrect functionality.

For simplicity, a combinational circuit C is assumed, with CNF representation CNF (C). The input stimulus I is represented as a set of assignments to the circuit inputs, $I = \{x_1 = v_1, \ldots, x_n = v_n\}$. The output responses are also represented as a set of assignments to the circuit outputs $O = \{z_1 = u_1, \ldots, z_m = u_m\}$. Finally, the actual circuit is represented by the CNF formula $CNF(C)$.

The unit clauses associated with I and O are hard clauses, i.e., must be satisfied, whereas the clauses associated with the circuit are soft clauses (i.e., may not be satisfied). The design debugging problem consists of satisfying the resulting CNF formula, such that the number of satisfied soft clauses is maximized, and all the hard clauses are also satisfied. Hence, the design debugging problem is naturally represented as an instance of the partial maximum satisfiability problem. [76]

8.4.4   Bounded model checking

Given a set of propositional symbols Σ, a Kripke structure is defined as a 4-tuple $M = (S, I, T, L)$, where S is a finite set of states, $I \subseteq S$ is a set of initial states, $T \subseteq S \times S$ is a transition relation, and $L : S \rightarrow P(\Sigma)$ is a labeling function, where $P(\Sigma)$ denotes the power set over the set of propositional symbols. Temporal logics allow the description of properties of systems. Two propositional temporal logics are widely used: linear-time logic (LTL) and computation-tree logic (CTL). [79] In this chapter, temporal properties are described in LTL, but CTL could also have been considered. Model checking algorithms can be characterized as explicit-state or implicit-state (or symbolic). [79] Explicit state-model checking algorithms represent the states of the transition relation explicitly, whereas symbolic model checking algorithms do not. Initial symbolic model checking algorithms were based on binary decision diagrams (BDDs). [8] Over the last decade, a number of alternatives based on Boolean satisfiability (SAT) have been proposed. [7–9]

Most work on SAT-based model checking assumes *safety* properties G ψ_s, where ψ_s is a purely propositional formula. The interpretation is that ψ_s must hold on *all* reachable states of M. For simplicity, the Kripke structure $M = (S, I, T, L)$ will be represented by the 3-tuple $M = (I, T, F)$, where I is a predicate representing the initial states, T is a predicate representing the transition relation, and F is a predicate representing the failing property (i.e., $F = \neg\psi_s$), defined on state variables (denoted as set Y). Moreover, the predicates I, T, or F assume the underlying Kripke structure $M = (S, I, T, L)$ and associated target formula ψ_s. Observe that the states are not explicitly represented. A set of variables Y encodes the possible states, and predicate T encodes whether the system can go from state (represented with variables) Y_i to state Y_{i+1}.

Algorithm 8.2 Or organization of BMC
BMC($M = (I, T, F), \mu$)
1 $k \leftarrow 0$
2 **while** $k \leq \mu$
3   **do** $\varphi \leftarrow$ CNF(BMC(M, k),W)
4   **if** SAT(φ)
5   **then return false** ▷ Found counter-example
6          $k \leftarrow k + 1$
7 **return true**

As mentioned earlier in the chapter, bounded model checking focuses on safety properties G ψ_s, denoting that ψ_s must hold globally. The solution to addressing this problem with SAT is to consider the complement $F \neg \psi_s$, representing the condition that ψ_s will not hold in some reachable state. The condition $\neg \psi_s$ will be referred to as the failing property, and represented with a predicate F. Bounded model checking consists of iteratively unfolding the transition relation, while checking whether the failing property holds. The generic Boolean formula associated with SAT-based BMC is: [7]

$$I(Y_0) \wedge \bigwedge_{0 \leq i < k} T(Y_i, Y_{i+1}) \wedge \left(\bigvee_{0 \leq i \leq k} F(Y_i) \right). \tag{8.17}$$

Equation (8.17) is referred to as BMC(M,k), and represents the unfolding of the transition relation for k time steps, where $I(Y_0)$ represents the initial state (at time step 0), $T(Y_i, Y_{i+1})$ represents the transition relation between states at time steps i and $i+1$, respectively Y_i and Y_{i+1}, and $F(Y_i)$ represents the failing property at time step i. Given the proposition formula BMC(M,k), it is straightforward to generate a CNF formula φ, as described earlier in this chapter. The resulting CNF formula can then be evaluated by a SAT solver.

The typical organization of BMC for safety properties is illustrated in Algorithm 8.2. The details regarding the sets of variables associated with each propositional formula are omitted, but are clear from the context. Moreover, the encoding of the BMC formula to CNF is shown as function CNF(), and uses a set of auxiliary Boolean variables W. Finally, μ represents an upper bound on the unfolding of the transition relation. Experimental evidence has confirmed SAT-based BMC to be an extremely competitive technique, which has been used in industrial settings. [81]

A key difficulty with BMC is its inability to prove that there is no counter-example for a given safety property G ψ_s. Unless the recurrence (or the reachability) diameter [81] of an automation is known, it is not possible to precompute the value of the upper bound (μ) used in Algorithm 8.2. In general, the recurrence diameter of an automaton is not known, and so BMC is incomplete. Hence, if the BMC algorithm returns true it does not imply that a counter-example cannot be identified. In recent years, different approaches have been proposed for ensuring the completeness of SAT-based model checking. These approaches will be referred to as *unbounded model checking* (UMC).

Well-known examples include the use of induction [8] and interpolation. [9] The next section outlines the use of induction.

8.4.5   Unbounded model checking

To describe UMC approaches, the following predicates are defined:

$$\text{UNFOLD}(M, r, s) = I(Y_r) \wedge \left(\bigwedge_{r \leq i < s} T(Y_i, Y_{i+1}) \right). \tag{8.18}$$

Equation 8.18 represents the unfolding of the transition system for $s - r$ time steps, with $s \geq r$. The first state (represented with state variables Y_r) must be one of the initial states. Each set of variables Y_i represents a state reached after $i - r$ time steps, starting from one of the initial states.

$$\text{TRAN}(M, s, t) = \bigwedge_{s \leq i < t} T(Y_i, Y_{i+1}). \tag{8.19}$$

Equation 8.19 captures the transition relation for $t - s$ time steps, with $t \geq s$.

$$\text{FAIL}(M, u, v) = \left(\bigwedge_{u \leq i < v} T(Y_i, Y_{i+1}) \right) \wedge \left(\bigvee_{u \leq i < v} F(Y_i) \right). \tag{8.20}$$

Equation 8.20 represents the transition relation for the last $v - u$ time steps, with $v \geq u$, during which the failing property is checked for.

Hence, we can express the BMC formula in terms of these predicates:

$$\begin{aligned} \text{BMC}(M, r, s, t) &= \text{UNFOLD}(M, r, s) \wedge \text{FAIL}(M, s, t) \\ &= \text{UNFOLD}(M, r, r) \wedge \text{TRAN}(M, r, s) \wedge \text{FAIL}(M, s, t). \end{aligned} \tag{8.21}$$

Sheeran *et al.* proposed the first complete approach for SAT-based UMC. [8] To present this UMC solution, let us introduce a predicate that holds true for paths with no repeated states in the transition system:

$$\text{LOOPFREE}(M, r, s) = \text{TRAN}(M, r, s) \wedge \bigwedge_{r \leq i < j \leq s} (Y_i \neq Y_j). \tag{8.22}$$

Algorithm 8.3 Induction-based UMC algorithm

$\text{UMC}(M = (I, T, F))$
1 $k \leftarrow 0$
2 **while true**
3   **do if not** $\text{Sat}(I(Y_0) \wedge \text{LoopFree}(M, 0, k))$ ▷ Check fixed point
4   **then return true**
5   **if not** $\text{Sat}(\text{LoopFree}(M, 0, k) \wedge \text{Fail}(M, k, k))$ ▷ Check fixed point
6   **then return true**
7   **if** $\text{Sat}(I(Y_0) \wedge \text{Tran}(M, 0, k) \wedge \text{Fail}(M, k, k))$
8   **then return false** ▷ Found counter-example
9      $k \leftarrow k + 1$

Algorithm 8.3 outlines the induction-based UMC algorithm of Sheeran *et al.* The existence of a counter-example is tested in line 7. Moreover, the induction step is tested in lines 3 and 5. If, for a given k, there can be no loop-free paths of length k starting from an initial state, and a counter-example has not yet been found, then a counter-example *cannot* be found. Similarly, if for a given k, there can be no loop free paths of length k reaching a failing property, and a counter-example has not yet been found, then a counter-example *cannot* be found. Further improvements to induction-based UMC, including the use of incremental SAT, are described in [82].

8.4.6   Other applications

Boolean satisfiability finds many other applications in EDA. Besides the applications described above, other well-known examples include verification of pipelined processors, [20,21] symbolic trajectory evaluation, [22] design debugging and diagnosis with SAT, [10] identification of functional dependencies in Boolean functions, [24] technology mapping in logic synthesis, [25] circuit-delay computation, [26] and cross-talk noise prediction [13]. The reader is referred to the bibliography for additional detail.

8.5   Conclusions

Boolean satisfiability is an NP-complete decision problem, and all existing algorithms require worst-case exponential time in the size of the problem representation. Nevertheless, modern SAT algorithms are remarkably efficient, capable of solving large, complex examples from real applications. The efficiency of SAT algorithms has motivated their use in an ever-increasing number of practical applications, ranging from cross-talk noise prediction in integrated circuits to termination analysis of term-rewrite systems, and including model checking of hardware and software systems. Moreover, SAT finds many natural applications in EDA, in the areas of verification, testing, and synthesis. This chapter summarizes the organization of the most effective SAT algorithms for solving practical EDA problems, and provides an overview of some of the most successful applications of SAT in EDA. Moreover, the chapter summarizes recent work on representative extensions of SAT, which are increasingly being used in EDA.

8.6   Acknowledgement

This work is partly supported by EU projects IST/033709 and ICT/217069 and by EPSRC grant EP/E012973/1.

8.7   References
[1] S. Cook (1971). The complexity of theorem proving procedures. In *Proceedings of the Third Annual Symposium on Theory of Computing*, pp. 151–158.

[2] J. Marques-Silva and K. Sakallah (1996). GRASP: a new search algorithm for satisfiability. In *International Conference on Computer-Aided Design*, pp. 220–227.

[3] M. Moskewicz, C. Madigan, Y. Zhao, L. Zhang, and S. Malik (2001). Engineering an efficient SAT solver. In *Design Automation Conference*, pp. 530–535.

[4] N. Een and N. Sörensson (2003). An extensible SAT solver. In *International Conference on Theory and Applications of Satisfiability Testing*, pp. 502–518.

[5] P. Chen and K. Keutzer (1999). Towards true crosstalk noise analysis. In *International Conference on Computer-Aided Design*, pp. 132–138.

[6] C. Fuhs, J. Giesl, A. Middeldorp, *et al.* (2007). SAT solving for termination analysis with polynomial interpretations. In *International Conference on Theory and Applications of Satisfiability Testing*, pp. 340–354.

[7] A. Biere, A. Cimatti, E. Clarke, and Y. Zhu. Symbolic model checking without BDDs (1999). In *Tools and Algorithms for the Construction and Analysis of Systems 5th International Conference*, pp. 193–207.

[8] M. Sheeran, S. Singh, and G. Stalmarck (2000). Checking safety properties using induction and a SAT solver. In *Formal Methods in Computer-Aided Design*, pp. 108–125.

[9] K. L. McMillan (2003). Interpolation and SAT-based model checking. In *Computer-Aided Verification, 15th International Conference*, pp. 1–13.

[10] A. Smith, A. G. Veneris, M. F. Ali, and A. Viglas (2005). Fault diagnosis and logic debugging using Boolean satisfiability. *IEEE Transactions on Computer-Aided Design*, **24** (10):1606–1621.

[11] B. Selman and H. Kautz (1992). Planning as satisfiability. In *European Conference on Artificial Intelligence*, pp. 359–363.

[12] J. Rintanen, K. Heljanko, and I. Niemela (2006). Planning as satisfiability: parallel plans and algorithms for plan search. *Artificial Intelligence*, **170**(12–13):1031–1080.

[13] I. Lynce and J. Marques-Silva (2006). Efficient haplotype inference with Boolean satisfiability. In *National Conference on Artificial Intelligence*.

[14] A. Darwiche (2004). New advances in compiling CNF into decomposable negation normal form. In *European Conference on Artificial Intelligence*, pp. 328–332.

[15] D. Jackson, I. Schechter, and I. Shlyakhter (2000). Alcoa: the Alloy constraint analyzer. In *International Conference on Software Engineering*, pp. 730–733.

[16] E. M. Clarke, D. Kroening, and F. Lerda (2004). A tool for checking ANSI-C programs. In *Tools and Algorithms for the Construction and Analysis of Systems*, pp. 168–176.

[17] S. Khurshid and D. Marinov (2004). TestEra: specification-based testing of java programs using SAT. *Automated Software Engineering Journal*, **11**(4):403–434.

[18] C. Tucker, D. Shuffelton, R. Jhala, and S. Lerner (2007). OPIUM: optimal package install/uninstall manager. In *International Conference on Software Engineering*, pp. 178–188.

[19] P. Manolios, M. G. Oms, and S. O. Valls (2007). Checking pedigree consistency with PCS. In *Tools and Algorithms for the Construction and Analysis of Systems. 13th International Conference*, pp. 339–342.

[20] M. N. Velev and R. E. Bryant. Effective use of Boolean satisfiability procedures in the formal verification of superscalar and vliw microprocessors. *Journal of Symbolic Computation*, **35**(2):73–106.

[21] P. Manolios and S. K. Srinivasan (2005). Refinement maps for efficient verification of processor models. In *Design, Automation and Testing in Europe Conference*, pp. 1304–1309.

[22] J.-W. Roorda and K. Claessen (2005). A new SAT-based algorithm for symbolic trajectory evaluation. In *Advanced Research Working Conference on Correct Hardware Design and Verification Methods*, pp. 238–253.

[23] T. Larrabee (1992). Test pattern generation using Boolean satisfiability. *IEEE Transactions on Computer-Aided Design*, **11**(1):4–15.

[24] C.-C. Lee, J.-H. R. Jiang, C.-Y. Huang, and A. Mishchenko (2007). Scalable exploration of functional dependency by interpolation and incremental SAT solving. In *International Conference on Computer-Aided Design*, pp. 227– 233.

[25] S. Safarpour, A. G. Veneris, G. Baeckler, and R. Yuan (2006). Efficient SAT-based Boolean matching for FPGA technology mapping. In *Design Automation Conference*, pp. 466–471.

[26] P. C. McGeer, A. Saldanha, P. R. Stephan, R. K. Brayton, and A. L. Sangiovanni-Vincentelli (1991). Timing analysis and delay-fault test generation using path-recursive functions. In *International Conference on Computer-Aided Design*, pp. 180–183.

[27] G. Audemard, P. Bertoli, A. Cimatti, A. Kornilowicz, and R. Sebastiani (2002). A SAT based approach for solving formulas over Boolean and linear mathematical propositions. In *International Conference on Automated Deduction*, pp. 195–210.

[28] H. Ganzinger, G. Hagen, R. Nieuwenhuis, A. Oliveras, and C. Tinelli (2004). DPLL(T): fast decision procedures. In *16th International Conference on Computer-Aided Verification*, pp. 175–188.

[29] V. Manquinho and J. Marques-Silva (2000). Search pruning conditions for Boolean optimization. In *European Conference on Artificial Intelligence*, pp. 130–107.

[30] N. Eén and N. Sörensson (2006). Translating pseudo-Boolean constraints into SAT. *Journal on Satisfiability, Boolean Modeling and Computation*, **2**:1–25.

[31] C. M. Li, F. Many'a, and J. Planes (2007). New inference rules for Max-SAT. *Journal of Artificial Intelligence Research*, **30**:321–359.

[32] F. Heras, J. Larrosa, and A. Oliveras (2007). MiniMaxSat: a new weighted Max-SAT solver. In *International Conference on Theory and Applications of Satisfiability Testing*, pp. 41–55.

[33] R. Bayardo, Jr. and J. Pehoushek (2000). Counting models using connected components. In *National Conference on Artificial Intelligence*, pp. 157–162.

[34] T. Sang, P. Beame, and H. A. Kautz (2005). Heuristics for fast exact model counting. In *International Conference on Theory and Applications of Satisfiability Testing*, pp. 226–240.

[35] R. Letz (2002). Lemma and model caching in decision procedures for quantified Boolean formulas. In U. Egly and C. G. Fermüller, eds., *International Conference on Automated Reasoning with Analytic Tableaux and Related Methods*, Lecture Notes in Computer Science, vol. 2381, pp. 160–175. Springer-Verlag.

[36] P. R. Stephan, R. K. Brayton, and A. L. Sangiovanni-Vincentelli (1996). Combinational test generation using satisfiability. *IEEE Transactions on Computer-Aided Design*, **15**(9):1167–1176.

[37] J. Marques-Silva and K. Sakallah (1997). Robust search algorithms for test pattern generation. In *IEEE Fault-Tolerant Computing Symposium*, pp. 152–161.

[38] M. Davis and H. Putnam (1960). A computing procedure for quantification theory. *Journal of the ACM*, **7**:201–215.

[39] R. Zabih and D. A. McAllester (1988). A rearrangement search strategy for determining propositional satisfiability. In *National Conference on Artificial Intelligence*, pp. 155–160.

[40] S. Subbarayan and D. K. Pradhan (2004). NiVER: non-increasing variable elimination resolution for preprocessing SAT instances. In *International Conference on Theory and Applications of Satisfiability Testing*, pp. 276–291.

[41] N. Eén and A. Biere (2005). Effective preprocessing in SAT through variable and clause elimination. In *International Conference in Theory and Applications of Satisfiability Testing*, pp. 61–75.

[42] G. S. Tseitin (1968). On the complexity of derivation in propositional calculus. *Studies in Constructive Mathematics and Mathematical Logic*, Seminars in Mathematics, vol. 8, part II, pp. 115–125. Steklov Mathematical Institute.

[43] D. A. Plaisted and S. Greenbaum (1986). A structure-preserving clause form translation. *Journal of Symbolic Computation*, **2**(3):293–304.

[44] P. A. Abdulla, P. Bjesse, and N. Een (2000). Symbolic reachability analysis based on SAT solvers. In S. Graf and M. Schwartzbach, eds., *Tools and Algorithms for the Construction and Analysis of Systems*, Lecture Notes in Computer Science, vol. 1785, pp. 411–425. Springer-Verlag.

[45] H. R. Andersen and H. Hulgaard (1997). Boolean expression diagrams. In *Twelfth Annual IEEE Symposium on Logic in Computer Science*, pp. 88–98.

[46] A. Kuehlmann, V. Paruthi, F. Krohm, and M. K. Ganai (2002). Robust Boolean reasoning for equivalence checking and functional property verification. *IEEE Transactions on Computer-Aided Design*, **21**(12):1377–1394.

[47] J. P. Warners (1998). A linear-time transformation of linear inequalities into conjunctive normal form. *Information Processing Letters*, **68**(2):63–69.

[48] O. Bailleux, Y. Boufkhad, and O. Roussel (2006). A translation of pseudo Boolean constraints to SAT. *Journal on Satisfiability, Boolean Modeling and Computation*, **2**:191–200.

[49] C. Sinz (2005). Towards an optimal CNF encoding of Boolean cardinality constraints. In *International Conference on Principles and Practice of Constraint Programming*, pp. 827–831.

[50] S. D. Prestwich (2007). Variable dependency in local search: prevention is better than cure. In *International Conference on Theory and Applications of Satisfiability Testing*, pp. 107–120.

[51] T. Walsh (2000). SAT v CSP. In *International Conference on Principles and Practice of Constraint Programming*, pp. 441–456.

[52] J. Gu, P. W. Purdom, J. Franco, and B. W. Wah (1997). Algorithms for the satisfiability (SAT) problem: a survey. In D. Du, J. Gu, and P. M. Pardalos, eds., *Satisfiability Problem: Theory and Applications*, DIMACS Series in Discrete Mathematics and Theoretical Computer Science, vol. 35, pp. 19–151. American Mathematical Society.

[53] M. Sheeran and G. Stalmarck (1998). A tutorial on Stalmarck's proof procedure for propositional logic. In *Proceedings of the International Conference on Formal Methods in Computer-Aided Design*.

[54] M. Davis, G. Logemann, and D. Loveland (1962). A machine program for theorem-proving. *Communications of the ACM*, **5**:394–397.

[55] B. Selman, H. Levesque, and D. Mitchell (1992). A new method for solving hard satisfiability problems. In *National Conference on Artificial Intelligence*, pp. 440–446.

[56] M. H. Schulz, E. Trischler, and T. M. Sarfert (1998). SOCRATES: a highly efficient automatic test pattern generation system. *IEEE Transactions on Computer-Aided Design*, **7**(1):126–137.

[57] P. Tafertshofer and A. Ganz (1999). SAT based ATPG using fast justification and propagation in the implication graph. In *International Conference on Computer-Aided Design*, pp. 139–146.

[58] W. Kunz and D. Pradhan (1994). Recursive learning: a new implication technique for efficient solutions to CAD problems-test, verification, and optimization. *IEEE Transactions on Computer-Aided Design*, **13**(9):1143–1158.

[59] R. E. Tarjan (1974). Finding dominators in directed graphs. *SIAM Journal on Computing*, **3**(1):62–89.

[60] C. P. Gomes, B. Selman, and H. Kautz (1998). Boosting combinatorial search through randomization. In *National Conference on Artificial Intelligence*, pp. 431–437.

[61] L. Baptista and J. Marques-Silva (2000). Using randomization and learning to solve hard real-world instances of satisfiability. In *International Conference on Principles and Practice of Constraint Programming*, pp. 489–494.

[62] E. Goldberg and Y. Novikov (2002). BerkMin: a fast and robust SAT-solver. In *Design, Automation and Testing in Europe Conference*, pp. 142–149.

[63] L. Zhang and S. Malik (2003). Validating SAT solvers using an independent resolution-based checker: practical implementations and other applications. In *Design, Automation and Testing in Europe Conference*, pp. 10880–10885.

[64] L. G. e Silva, L. Silveira, and J. Marques-Silva (1999). Algorithms for solving Boolean satisfiability in combinational circuits. In *Design, Automation and Test in Europe Conference*, pp. 526–530.

[65] A. Kuehlmann, M. Ganai, and V. Paruthi (2001). Circuit-based Boolean reasoning. In *Design Automation Conference*.

[66] A. Mishchenko, S. Chatterjee, R. K. Brayton, and N. Eén (2006). Improvements to combinational equivalence checking. In *International Conference on Computer-Aided Design*, pp. 836–843.

[67] Z. Fu, Y. Yu, and S. Malik (2005). Considering circuit observability don't cares in CNF satisfiability. In *Design, Automation and Testing in Europe Conference*, pp. 1108–1113.

[68] J. Marques-Silva (2000). Algebraic simplification techniques for propositional satisfiability. In *International Conference on Principles and Practice of Constraint Programming*, pp. 537–542.

[69] F. Bacchus (2002). Enhancing Davis Putnam with extended binary clause reasoning. In *National Conference on Artificial Intelligence*.

[70] I. Lynce and J. Marques-Silva (2003). Probing-based preprocessing techniques for propositional satisfiability. In *International Conference on Tools with Artificial Intelligence*, pp. 105–110.

[71] P. Barth (1995). A Davis–Putnam enumeration algorithm for linear pseudo-Boolean optimization. Technical report MPI-I-95-2-003, Max Planck Institute for Computer Science.

[72] Z. Fu and S. Malik (2006). On solving the partial MAX-SAT problem. In *International Conference on Theory and Applications of Satisfiability Testing*, pp. 252–265.

[73] J. Marques-Silva and J. Planes (2008). Algorithms for maximum satisfiability using unsatisfiable cores. In *Design, Automation and Testing in Europe Conference*.

[74] N. Dershowitz, Z. Hanna, and J. Katz (2005). Bounded model checking with QBF. In *International Conference on Theory and Applications of Satisfiability Testing*, pp. 408–414.

[75] D. Brand (1993). Verification of large synthesized designs. In *International Conference on Computer-Aided Design*, pp. 534–537.

[76] S. Safarpour, H. Mangassarian, A. Veneris, M. H. Liffiton, and K. A. Sakallah (2007). Improved design debugging using maximum satisfiability. In *Formal Methods in Computer-Aided Design, 7th International Conference*, pp. 13–19.

[77] J. Marques-Silva and T. Glass (1999). Combinational equivalence checking using satisfiability and recursive learning. In *Design, Automation and Test in Europe Conference*, pp. 145–149.

[78] M. Abramovici, M. A. Breuer, and A. D. Friedman (1990). *Digital Systems Testing and Testable Design*. Computer Science Press.

[79] E. M. Clarke, O. Grumberg, and A. Peled (1999). *Model Checking*. MIT Press.

[80] K. L. McMillan (1993). *Symbolic Model Checking*. Kluwer Academic.

[81] A. Biere, A. Cimatti, E. Clarke, O. Strichman, and Y. Zhu (2003). Bounded model checking. *Advances in Computers*, **58**:118–149.

[82] N. Een and N. Sorensson (2003). Temporal induction by incremental SAT solving. In *Workshop on Bounded Model Checking*, ENTCS, vol. 89.

Index